THE FIRST STARGAZERS

THE
FIRST
STARGAZERS

An Introduction to the Origins of Astronomy

JAMES CORNELL

LONDON
THE ATHLONE PRESS
1981

To my parents

All photographs and illustrations are by the author unless otherwise specified.

THE ATHLONE PRESS,
90-91 Great Russell St.
London WC1B 3PY England

Printed in the United States of America

ISBN: 0 485 30004 4

Contents

Acknowledgments

Although the field of archaeoastronomy is relatively young, a wealth of information already has appeared in print concerning the attempts by modern scientists to trace the roots of science through the clues left in the art and artifacts of prehistoric and preliterate societies. As I am neither a professional archaeologist nor a professional astronomer, but one who has written about both fields, my intention in writing this book was not to break new ground with original research or even to pose new theories about how ancient man used the heavens. Rather, I have tried to summarize, or, more precisely, synthesize much of the voluminous, but somewhat scattered, material into a single comprehensive overview.

Because the study of archaeoastronomy is still growing—and even rapidly expanding—this volume is also designed as an introduction for general readers, with the hope that they will be tempted by my broad approach to explore some of the many different avenues taken by specialists in the past two decades. Fortunately, if readers do pursue the subject further, they will find many excellent guides. In fact, one of the most disarming features of archaeoastronomy for the popular-science writer is how well some of its prime movers do themselves write. The series of volumes on New World astronomy edited by Anthony Aveni, the lively *Archaeoastronomy Bulletin* issued by John B. Carlson, the Polynesian sagas of David Lewis, Gerald Hawkins's *Stonehenge Decoded* and *Beyond Stonehenge*, as well as two excellent anthologies— *In Search of Ancient Astronomies*, edited by E. C. Krupp, and *Astronomy of the Ancients*, edited by Kenneth Brecher and Michael Fiertag—all demonstrate how clearly, colorfully, and enthusiastically some scientists can communicate about their work. I owe a large debt to these and other equally articulate scientists and writers in the field and I only hope I have expressed some of the concepts as completely and accurately as they have already done.

I am especially grateful to five such authors—John B. White, Owen Gingerich, Kenneth Brecher, John Eddy, and Gerald Hawkins—for reading my manuscript in draft and making many constructive comments on its contents. Ronald R. La Count also read the manuscript and made many contributions throughout its long gestation period, sending me a steady flow of suggestions and new lines of inquiry that served to spur me on whenever my fervor flagged.

At Scribners, Doe Coover encouraged me to do the book and was responsible for starting the project; later, Dwight Allen and Nancy Palmquist suffered through the pains of the editorial process; and, in the waning days, Charles Scribner, Jr., helped me over the last remaining rough spots.

Obviously, none of the above should be blamed for any errors, misinterpretations, inconsistencies, or, as it may appear to some specialists, oversimplifications of very complex concepts. All of that is entirely my own doing.

I should also give general thanks to the host of individuals who contributed special pieces of information, sometimes directly, as in the case of Travis Hudson and Von Del Chamberlain, but more often indirectly through interview or simple assimilation at various press conferences and scientific meetings. My special thanks, too, to two good friends in Mexico: Guillermo Aldana, who showed me how to look at the play of light and shadow on ancient monuments with a photographer's eye; and Gabriel Muñoz, who introduced me first to ancient Tarascan astronomy and later to the delights of modern Michoacán.

The illustrations came from a variety of sources, to all of whom I am most grateful, including William Hauser of the United States Department of Agriculture Forest Service, Bruce Hopkins of the National Parks Service, Tom Harney of the Smithsonian Institution's Museum of Natural History, James Glenn of the Smithsonian's National Anthropological Archives, Von Del Chamberlain of the Smithsonian's National Air and Space Museum, Charles A. Federer of *Sky and Telescope*, and Gerald Hawkins. Additional illustrations were generously provided by the Mexican Government Tourism Office and the British Tourist Authority, as well as the Harvard-Smithsonian Center for Astrophysics (these include the astronomical schematics originally prepared by John Hamwey, Joseph Singarella, and Charles Hanson for a public presentation on this same subject). Mary Juliano had the unenviable task of deciphering and typing my nearly illegible notes and early drafts—and then retyping what seemed like endless rewrites.

Finally, and most important, thanks to my wife and daughter for their companionship on all the visits to far-flung archaeoastronomy sites and for their understanding and patience during the many hours required later to put the record of our travels on paper.

Figure 1. *The pyramid of Kukulcán at Chichén-Itzá, Yucatan, Mexico. At sunset on the equinoxes, an intricate interplay of sun and stone creates the image of a giant serpent down the staircase of this stepped pyramid.*

Introduction

> *"The intellectual activity of mankind during prehistory is*
> *a vast almost uncharted ocean. . . . There have been only*
> *about 200 generations of* history *[but] upwards of 10,000*
> *generations of* prehistory. *. . . Among the great throng, it*
> *seems to me likely that some must have gazed up at the sky*
> *and wondered earnestly about the sun, moon, and the stars.*
> *They would have done so with a basic intelligence equal to*
> *our own. . . ."*
>
> —Fred Hoyle, *On Stonehenge*

The pyramid of Kukulcán at Chichén-Itzá looms over the low
scrub jungle of the Yucatan Peninsula like some misplaced medieval fortress. Indeed, the conquistadors, the first Europeans to see
this magnificent monument, called it *El Castillo*, the castle. Eighty
feet high and 183 feet on a side, the pyramid is actually a series of
nine stepped platforms topped by a small square temple that is
reached by steep staircases. On the two sides that have been fully
restored, the staircase walls take the forms of elongated and
stylized snakes—the plumed serpent of Mesoamerican mythology—each terminating at the foot of the steps in a grotesque
reptilian head four-and-a-half feet high. Even today, with a highway running a few yards away and a shabby collection of taco
stands within sight, the pyramid of Kukulcán is awesome (see
Figure 1).

1

At dusk on either the vernal or autumnal equinox, when the sun sets almost due west, visitors to this Mayan ruin may see a phenomenon that makes the pyramid even more impressive—and intriguing. As the sun drops toward the featureless horizon, its slanting rays strike the corners of the nine stepped platforms so that a saw-toothed pattern of light and shadow is formed across the wall of the staircase on the northern façade. Seven triangles of light appear in sequence, slowly descending the serpent-body bannister from the top of the pyramid to the gaping mouth of the serpent's head. For one moment, when the sun reaches the proper angle, the entire serpent is painted with light so its stone body seems to have the distinctive diamondback markings of a rattler.

Then, as the sun sinks lower, the shadow of the night rises from the ground and the triangles of light are extinguished one by one, beginning with the serpent's head and progressing back up the body. Finally, the last triangle at the very top of the pyramid vanishes and the entire structure is engulfed in darkness.

Is it merely coincidence that sun, shadow, and stone work out this brief spectacle at Chichén-Itzá on the equinoxes? Not likely, when one considers that the vernal equinox, usually occurring around March 21, signals for the people of the Yucatan—today as it did in the twelfth century A.D., when the pyramid was built—the time to clear and burn off the forest for planting in preparation for the rainy season. Conversely, six months later, the autumnal equinox (around September 21) marks the end of the summer rains. The ancient Mayan calendar marked the March equinox as an important date; and even after the arrival of the Spanish, Bishop Diego de Landa noted a festival held in the month of Xul to honor Kukulcán and the end of the year.

Luís Enrique Aroche, who has recorded the equinox phenomenon at Chichén-Itzá, suggests the descending pattern of light may have been intended to represent the return of Kukulcán from heaven. Offerings might have been placed at the mouth of the serpent, itself a symbol of this deity who, like the Son of God in Christian theology, died and was born again. It is interesting, too, that inside the pyramid, just below the temple and about at the point where the last triangle of light disappears, there is a

2

secret chamber enshrining a bejeweled statue of the sacred jaguar.

No matter if its purpose was religious (to celebrate the rebirth of a god) or calendric (to herald important agricultural dates), the design of the pyramid to create this light show at the equinoxes seems intentional. And to achieve this intricate interplay of sun and shadow required extraordinary engineering skills, a familiarity with basic geometry, and an awareness of solar motion. In short, the Mayan builders of El Castillo must have had a knowledge of observational astronomy and utilized it to determine their architecture, their cultural celebrations, and their economy.

The Maya were not unique in this ability. During the past decade, increasing evidence has been found to suggest that almost all ancient peoples had not only an awareness but an understanding of celestial phenomena. In areas as disparate as northern Scotland and subsaharan Africa and as widely separated in time as Pharaonic Egypt and pre-Hispanic Mexico, a surprising number of so-called prescientific cultures had developed a sophisticated grasp of basic astronomical principles.

Beyond merely observing and recording the sun, moon, and stars as objects of wonder, many of these early societies apparently understood the cyclic nature of certain celestial phenomena and used the knowledge to create calendric devices that signaled important dates in their cultures. In some cases, supposed "primitives" may even have mastered the ability to predict eclipses and approached an understanding of the earth's rotundity.

Consciously or unconsciously, this knowledge of the heavens, and particularly the cyclic nature of certain events, helped these societies adapt to their environment and establish stable, permanent settlements. For ancient man, the discovery of the stars may have been the first step toward scientific, or at least cognitive, awareness, thus laying the foundations of modern civilization. The records of these fledgling efforts in astronomy are found today in the stone monuments these peoples left behind.

The deciphering of these incomplete, uneven, and often inconclusive records is the work of a new discipline of science, archaeoastronomy. According to the semiofficial journal, the *Archaeoastronomy Bulletin*, "Archaeoastronomy is the study of the astronomical practices, cosmological systems and celestial lore of ancient and pre-technical peoples by means of their material legacy." Many other descriptions are possible—and legitimate. Indeed, the field is so young its goals and guidelines are still fluid, and it is even called by several names, including astroarchaeology. The research requires the merger of astronomy, archaeology, anthropology, ethnology, and history, while also depending on contributions from mythology, geometry, geography, ecology, surveying, folklore, and cryptology. It is obviously not the neat, homogeneous commingling of common backgrounds and similar talents favored by most sciences.

Out of this strange mixture of approaches and viewpoints is emerging a new understanding and appreciation of our prehistoric ancestors. In fact, archaeoastronomy has, in part, helped revolutionize our assessment of ancient man's intellectual and cultural development. Since the beginning of recorded history, all societies have been fascinated with the remains of those cultures that preceded them. In the fifth century B.C., Herodotus could be thrilled by the great monuments along the Nile, already over two thousand years old. The masterworks of antiquity—the pyramids of Egypt, the temples of Mesoamerica, the stone circles of the British Isles— have all inspired wonder, awe, and speculation about their purposes.

Obviously, many sites were tombs, ritual monuments, and ceremonial meeting grounds, but some observers also considered whether these ruins might have had other, practical purposes as well. One of the most persistent theories has been that many stone monuments were used as astronomical observatories.

For some monuments such astronomical relationships are supported by inscriptions, legends, and folklore linking the structures with a particular astral deity, or by obvious alignments with the cardinal points, or by an observed orientation toward the rising and setting sun on a specific date. For other structures, the relationship is more subtle. However, as Jerome Lettvin has noted, "astronomers, like other craftsmen, recognize the tools of their trade." No matter how crude or cumbersome they might seem, certain ruins have the uncanny appearance of observational instruments.

4

For example, in 1890, while traveling in Egypt, Sir Norman Lockyer noted that the long, straight peristyle corridor of the temple at Karnak, with its many side openings, or apertures, looked remarkably similar to an optical instrument known as a collimator used in modern telescopes (see Figure 2). Pursuing this point,

Figure 2. *The peristyle hall at Karnak, Luxor, Egypt. Lockyer saw the columned hall as resembling an astronomical instrument.*

Lockyer proposed that the corridor at the time of the summer solstice would allow the sun's rays to penetrate deep into the temple's inner chamber and illuminate the statue of Ra, the sun god.

In 1894, Lockyer published *The Dawn of Astronomy*, in which he proposed that Karnak had been deliberately and precisely laid out in accordance with astronomical principles. He also described other sites, including Stonehenge, that had astronomical significance. Although Lockyer was a distinguished scientist—he had first identified the element helium in the sun's chromosphere and was the founder and long-time editor of the British scientific journal *Nature*—his theories were not readily accepted by the scientific community. Many colleagues felt he had ventured too far from his own specialty and that his speculations were merely the eccentricities of an aging scholar.

In fact Lockyer was on the right track, and many of his insights would form the foundation for modern investigations. Unfortunately, he often allowed his speculations to far exceed the observed evidence. In particular he attempted to date the construction of monuments on the basis of the risings and settings of certain stars he assumed the buildings had been erected to observe. If the main axis of a monument aligned with the rising point of a bright star in 2000 B.C., Lockyer assumed the monument was constructed in that year. (The reverse procedure is applied today by most conscientious researchers—that is, they obtain the best date for the probable construction through other archaeological techniques and then attempt to find alignments with prominent celestial objects during that epoch.)

Alas, it was the more speculative nature of Lockyer's work that both made his book popular and gave it such ill repute in the scientific community. The imitators who followed him, often exceeding even Lockyer's wildest theories, served to cast further doubt on the basic approach of archaeoastronomy and made it one of the most controversial fields in science.

Part of the problem was that archaeoastronomy (particularly the sometimes questionable excursions into such areas as "pyramidology") was primarily the pursuit of inspired amateurs without academic affiliations. Even when studied in the context of formal

Figure 3. *Stonehenge at dawn.* Photo courtesy British Tourist Authority

scholarship, archaeoastronomy had no well-defined roots in either the physical or the social sciences; at best it hovered on the fringes of the history of science. Thus, in Mesoamerican archaeology, where the evidence for astronomical achievement was quite obvious, it was either ignored or considered peripheral to the study of the culture in general. Worse yet, many hypotheses were propounded by astronomers who could not be expected to be fully schooled in archaeological techniques, and, in turn, attacked by archaeologists unfamiliar with astronomy or mathematics. Until the mid-1970's, astronomers and archaeologists would remain at odds over the interpretation of even the most obvious astronomical records of the past.

In 1963, however, a short scientific paper appeared that, while initially fanning more controversy, eventually led to the interdisciplinary approach of current archaeoastronomy. That paper was Gerald S. Hawkins's computer-derived interpretation of Stonehenge as an astronomical observatory (see Figure 3). Hawkins showed that postholes, stones, and archways at Stonehenge lined up with sun and moon. In a second paper, he demonstrated that the builders might have known about eclipses and perhaps used the various circles of stones and holes as a counting device, or crude Neolithic computer. Other researchers, including Lockyer, had ascribed an astronomical function to Stonehenge (it had also been described as a variety of other things: from giant fairy ring to

7

Druid temple to prehistoric storage battery!), but Hawkins's explanation at least had the apparent impartiality and assumed certainty of modern electronic data-processing behind it. The public eagerly accepted the explanation as did many American archaeologists and anthropologists. By contrast, many British classical archaeologists, a group traditionally neither familiar nor comfortable with computers or fundamental astronomy, were hesitant to believe Hawkins's conclusions.

At the heart of the controversy was the apparent incompatibility of the sophistication of the Stonehenge "observatory" with the relatively low intellectual level of its supposed creators as revealed by other artifacts of their culture. More important, perhaps, Hawkins's finding had upset the timetable of human development, turning back the clock nearly two thousand years.

The Stonehenge debate, which continues today and about which more will be said later, served one valuable end. It brought general attention to a subject long ignored by scientists and misunderstood by the public. Most of the serious literature had been written within the narrow context of an author's own specialty. And on the popular level most treatments had been either pseudoscientific or sensational, in the mode of Piazzi Smyth or Erick Von Däniken, who found prophetic powers in the pyramids of Egypt or claimed ancestral connections between monument makers and extraterrestrial visitors.

By the early 1970s, archaeoastronomy, both as a theory and as a discipline, had finally won a grudging acceptance in the scientific world. Its first major conference, "The Uses of Astronomy in the Ancient World," was sponsored by the British Academy and the Royal Society of London in 1971 to bring the disparate disciplines together and, perhaps, to relieve the tensions created by the Stonehenge controversy. More important, a conference sponsored by these prestigious societies clearly stated that archaeoastronomy was worthy of attention and recognition as a valid field of research.

A second major international meeting, this time concentrating on pre-Columbian astronomy, was held in June 1973 at Mexico City under the joint sponsorship of the American Association for the Advancement of Science (AAAS) and the Consejo Nacional de

Ciencia y Tecnología of Mexico. As noted in the proceedings of this meeting, published later as the volume *Archaeoastronomy in Pre-Columbian America*, the organizers had hoped the conference would "legitimatize the field and promote cooperation between anthropologists, astronomers, and historians of science to reveal much about preliterate societies' awareness of the environment." Without doubt, the effort was a success, for in the United States and abroad in the next few years a series of similar meetings was held. Most notable, perhaps, was a conference at Colgate University in 1975 that resulted in a book, *Native American Astronomy*, and a three-day conference at the Massachusetts Institute of Technology in early 1977, the proceedings of which later appeared as a special issue of *Technology Review* and eventually as another book, *Astronomy of the Ancients*.

Also in 1977, the Center for Archaeoastronomy was established at the University of Maryland and began publishing the *Archaeoastronomy Bulletin*. Since 1979, the British-based *Journal of the History of Astronomy* has published a special supplement devoted solely to archaeoastronomy. And, in early 1980, following a special one-day session devoted to archaeoastronomy at the annual meeting of the AAAS, the American Astronomical Society decided to include the field as part of the Historical Astronomy division.

More important than merely sparking meetings or learned journals, the emergence of archaeoastronomy as a specific discipline has provided both the impetus for cooperative efforts between different scientific specialties and the inspiration for increased field research. In less than a decade, the catalog of astronomically related sites has been expanded from a few scattered ruins in Britain, Egypt, and the Yucatan into literally scores around the world.

Besides Stonehenge, dozens of other elaborate stone rings in the British Isles and northwestern Europe have been studied and found to have definite astronomical alignments, primarily with the solstices. Several burial mounds and long barrows also appear to have astronomical significance.

A "mini-Stonehenge" has been found at Namoratunga in northwest Kenya. Here an arrangement of nineteen basalt pillars is oriented toward certain stars and constellations, some of which are still used by modern peoples to mark dates in their calendar. Since this site was erected around 300 B.C., it suggests that the prehistoric peoples of East Africa developed a calendar based on astronomical knowledge.

In Mesoamerica, where an astronomical awareness has been long recognized, new evidence is emerging to suggest that the Mayas, in addition to developing an extremely accurate calendar and calculating the length of the year to an accuracy of a few decimal points, may have been close to understanding that Venus orbits the sun and determining its revolution period.

Astronomical knowledge appears to have been widespread throughout Mesoamerica, and a more accurate record of the spread of culture and transfer of technology from one society to another may be possible by tracing the astronomical traditions. Indeed, some of those traditions may have diffused northward to the North American Indians.

Once thought to be no more than loutish nomads who lived off the land as hunters and gatherers, the native Americans are now being reconstructed as complex, intelligent peoples who developed stable and intricate societies relatively advanced within the limits imposed by their environment. Part of this new image is due to research in archaeoastronomy.

In the Ohio and Mississippi valleys, great earthen mounds previously considered no more than huge cabalistic symbols—or at best graves and fortresses—now are found in at least one case to be astronomically aligned. Some show direct influences of Mesoamerican building techniques; others display a precision in design and layout that implies considerable engineering ability and a thorough knowledge of geometry. At Cahokia, Illinois, one of the largest mound complexes in the United States and a site that housed over twenty-five thousand people, researchers have found what appears to be an "American Woodhenge." A giant ring of holes marks what was once a circle of tall posts. By standing slightly off center in the enclosed circle, an observer can mark the direction of the solstices with extreme accuracy. Similar structures (on a smaller scale) have been found in Kansas; and several Plains Indian tribes apparently constructed lodges in the same fashion, leaving the doorways open and aligned with some significant point on the horizon. Farther west, in the high plains and mountain plateaus on the eastern slopes of the Rocky Mountains, several stone rings, or "medicine wheels," have been discovered with clear relationships between their layouts and, again, the solstice sunrise directions.

Many other examples of recent discoveries could be given. Indeed, this is a book about discovery: how ancient man discovered the sky; how archaeoastronomers, in turn, are discovering the unwritten record of the early scientific achievements; and, finally, how this record may help all of us discover something about ourselves.

The real contribution of the archaeoastronomers has been to force a reevaluation of our primitive ancestors. Those ancient men and women not only looked very much like us physically, but their brains must have been very similar, too. While they lacked the advantages of writing and recording, and therefore didn't have formal repositories of knowledge as we know them, they could store facts and figures in their collective memories. In addition, they were certainly capable of systematically observing the skies. They could even construct instruments for doing so, although the equipment might be rather large and cumbersome.

It does not matter if the purpose of these observations was merely to plan religious ceremonies or exorcise demons or appease some god. The effort still required intelligence, skill, and, for most large sites, incredible community cooperation. We still don't know if the observations of the stars, sun, and moon cycles led to an awareness of other natural phenomena, such as seasonal changes, animal migrations, and the growing habits of edible plants; or if just the reverse was true, and the observation of these natural and vital cycles led to a desire to predict them more accurately with a sky calendar. The intellectual development may not be as important as the simple awakening to the cyclic nature of life and the need for counting, measuring, and time-dating skills.

Most remarkable among archaeoastronomy's revelations is the universality of this intellectual awakening in mankind. Both interest in the sky and skills in astronomy seem to have developed independently and naturally on different continents in different eras among peoples as unrelated in time and character as the builders of Stonehenge and the cliff dwellers of the American Southwest. Some attempts at astronomy dead-ended, others flowered into sophisticated cosmologies, and still others borrowed concepts and techniques from contemporary cultures. But all seem intended as a response to the human desire to fix man's place in the universe, to control the vast and frightening environment by understanding it.

The widespread—and in many cases apparently spontaneous—emergence of astronomical awareness would seem to refute those claims that all knowledge originated in one cradle of civilization, be it Egypt or Mesopotamia, and then diffused slowly around the world. Similarly, the relatively early emergence in pretechnical societies of often sophisticated skills in geometry, mathematics, and astronomy demonstrates that humankind needed no help from "outer space" to get science started. No men from Mars or "gods" from another galaxy came with packaged information on temple building or star sighting. Rather, our skills and knowledge came naturally from ourselves, developed gradually through tedious observations and painful trial and error, and improved with practice over many generations.

But if the recent acceptance of archaeoastronomy as a science has not come easily, neither has the realization that the foundations of civilization are to be found somewhere in the dim reaches of prehistory. In a sense, then, this book is also about the bonds between the astronomers of today and those of yesterday.

Discovering
the Ancient Sky

"Before the skill of writing came the skill of remembering."
—Jerome Lettvin

The Majes Valley of Peru's Castilla Province slices through the barren desert some three hundred miles south of Lima to create a thin green oasis. Unlike the other deep coastal valleys, where the rivers flowing down the western slopes of the Andes have been reduced to mere trickles in their struggle to reach the sea, the water runs strongly here, providing irrigation for several hundred acres of farmland. (The river also abounds in *langostinos*, the freshwater crustaceans that are a great source of local pride and delight.)

It was in valleys such as this that the first civilizations of South America took root. Bands of nomadic hunters, relentlessly pushing southward thousands of years after their ancestors had crossed the Bering land bridge, settled here. Initially they were loose clans of hunters who turned farmers, but these people would create a series of widely separated mini-societies that would rival in art and industry any other preliterate cultures.

Today a rough single-lane dirt track runs from the river bank through green fields and then up the steep canyon slope until it dead-ends at a place known as Toro Muerto. Even Land Rovers can go no farther than this, and visitors must proceed by foot.

The sun is blinding here above the valley haze, and one is only dimly aware of being surrounded by a vast field of angular boulders. Thousands of rocks, ranging in size from footballs to small houses, are scattered for miles along the sandy slope paralleling the river below. As the shapes emerge from the glare, they are transformed into an open-air gallery.

Every available flat surface on almost every rock is emblazoned with figures: lines and curlicues, flowers and fishes, vines and tendrils, animals both familiar and fantastic.

Approximately one thousand years ago, Toro Muerto (a modern name, meaning "dead bull") was a ceremonial center for the Huaris, a pre-Incan people who inhabited this valley and the hills above it. And on these rocks they carved a vivid record of their lives and beliefs.

The boulders are soft sandstone, and any sharp object, even a piece of hard wood, can easily scrape the dark-brown outer surface to expose a lighter shade below. Because it seldom rains in Peru's coastal desert, the rocks age and erode very slowly, and the markings scraped on their surfaces remain as sharp and clear today as when the first travelers passed this way a millennium ago.

To the casual visitor the rock carvings appear only as simple designs in a style common in other pre-Columbian art. Yet most are really stylized and simplified drawings of common objects, flora and fauna, natural phenomena, and scenes from Huari daily life. Perhaps the carvings had symbolic meaning for the Huaris, but they also accurately recorded their times. Professional botanists, biologists, and ethnologists can easily spot representations of familiar subjects: llamas cavort across the altiplano, hunters stalk birds and pumas, fish swim in great rivers, herdsmen drive stock into crude corrals, and masked dancers form ancient chorus lines.

Not surprisingly, then, astronomers can also recognize familiar elements in the scene carved on one huge boulder. A long, two-headed serpent winds its way across the rock face through a field of open circles. One circle has a smaller concentric ring within it and lines radiating from its perimeter. A sun symbol, perhaps? Similarly, the small crescent may represent a phase of the moon. A humanoid figure appears at one edge beside an indefinite animal shape. The more easily recognized images of two llamas—one thin, the other full-bodied—are carved above and below the sun (see Figure 4, page 16).

Professor Eloy Linares Málaga, an archaeologist specializing in petroglyphs, who led me to this stone, interprets the carving as simply a representation of natural celestial phenomena that the Huaris observed and recorded. The two-headed serpent could be the dense swarm of stars in the Milky Way that crosses the sky dome overhead, with the double head symbolizing the visible dark lane that splits it. The sun and crescent moons are even more literal representations. And the other smaller circles, while possibly bright stars, are most likely moon symbols marking the year's division into twelve lunar months. This calendrical quality of the carving is further strengthened by the "fat" and "lean" llamas, which may represent summer and winter, the only two seasons experienced at this latitude.

However, when I later showed a photo of the rock carving to an archaeoastronomer in the United States, he felt that the serpent was not the Milky Way, but rather the ecliptic—the apparent path of the sun through the stars as seen from earth. The disposition of the sun and the moon circles near the serpent suggest this. Moreover, the fact that moons are found on either side of the serpent (five above and seven below, if the crescent is included) suggests also that the carver had an awareness that the moon, because its orbit is tilted in respect to the ecliptic, each month goes north and then south of this imaginary line. The waviness of the serpent might be an abstract design representing the moon's apparent motion. The eight complete oscillations in the serpent are more difficult to explain, but Huaris might well have been tracing some

Figure 4. *An eleven-hundred-year-old petroglyph from Toro Muerto in Peru shows a simple astronomical design that combines a large sunburst with twelve lunar symbols representing the year, and both fat and lean llamas representing the two seasons. The two-headed serpent running through the moons has been interpreted as the Milky Way.*

longer cycle related to the prediction of solar eclipses, since solar eclipses occur only when the sun passes through the nodes, the intersections of the moon's orbit with the ecliptic.

These two interpretations of the Huari stone carving at Toro Muerto typify the two, sometimes divergent, views of modern science toward the astronomical awareness of ancient peoples. The first view, traditional among archaeologists, is simply that prehistoric peoples saw the sky and recorded what they saw. The second view suggests that ancient peoples not only were aware of the intricate motions of celestial bodies, but also recognized the cyclic nature of certain phenomena and even made attempts to predict future events. At the very least they made direct connections between the solar cycles and seasonal changes.

Just as with many other remains of preliterate cultures, the message on the rock of Toro Muerto is far from conclusive. One's interpretation of the record depends on one's own scientific—and personal—bias. Yet it is unquestionably clear that the Huaris, like other early peoples, did observe the sky and were familiar with the objects seen there. Less clear is exactly when they began to make connections between the cyclic nature of celestial events and the more mundane phenomena of daily life.

In today's world, most of us live under an almost permanent blanket of atmospheric pollution produced by the effluvia of internal-combustion engines, industrial furnaces, and open fires. The lights of an urbanized world bounce off these suspended particles to create a milky overcast on even the clearest nights. Waves of neon and mercury-vapor light now lap at the edges of our largest telescopes, making all observations difficult—and in some wavelengths impossible. To escape this light pollution, professional astronomers must seek out even more remote locations for their instruments such as the mountaintops of Chile or extinct volcanoes in the Pacific.

The casual observer has also lost the splendor of the nighttime sky behind the haze of modern pollution. (To judge how unfamiliar we have become with the heavens, visit a planetarium and listen to the gasps of surprise when the dome is suddenly filled with thousands of stars. For most children—and even many adults—it's a sight never seen before. Indeed, many planetarium goers often remark on how "unnatural" the sky show seems.) More important, modern man no longer needs the sky. We have digital alarms to ring out the hours. Calendars tell us days and years. Advertising and public-relations people remind us of special holidays and coax us into seasonal behavior. Compasses and atlases provide direction and precise mileage for our journeys. Observatories predict eclipses and occultations, and television turns these natural phenomena into media events. While the sky may serve some aesthetic purpose and provides the backdrop to our leisure activities, it no longer has much practical value for us. Only a handful of professionals—astronomers, meteorologists, navigators—ever seriously study the heavens.

Not so for ancient man. In primitive societies, the sky was map, calendar, clock, and more—more, perhaps, than modern humans can ever conceive. The sky was both an integral part of daily life and a presence of supernatural power. The sun and moon were beings with character, personality, and gender. The stars, as they rolled slowly from horizon to horizon, became animate objects imbued with spirit and intelligence. The celestial bodies inspired awe and admiration, but knowledge of them also had practical applications.

Consider the ancient hunter leaving his encampment in the morning to forage for food. He needed some guide as to the distance he could wander and still return before nightfall. The sun provided a perfect clock; if he traveled until the sun was directly overhead, he would have to turn back at this time to reach home by dusk. The sun was also a directional beacon, with its rising and setting points east and west providing two vital signposts. At night the stars too could provide directions, for although they rise in the east and set in the west, they also seem to turn about a point in the north where one star remains stationary. If bright enough to attract attention, that star could, as Polaris does for us today, serve as a constant indicator of the other two cardinal points of the compass. The slow motion of the other stars around it would serve as a crude clock. Moreover, different clusters of stars would appear above the horizon at different times of the year, thus marking changes in

18

weather and signaling the time for certain activities, such as gathering fruits or preparing fur skins for the winter.

The relationship between the sun and the seasons would have been obvious for peoples living in the northern hemisphere; the sun would be high in the sky—nearly overhead—at midday in summer and noticeably lower in the sky at midday in winter. Similarly, the point where the sun rose on the eastern horizon (and correspondingly set on the western horizon) would change with the seasons, rising farther and farther north in the warm summer months and farther and farther south in the cold winter months. The extreme points of sunrise and sunset would have been observed and noted as important, for they would signal times of consequence to hunting and gathering peoples who lived exposed to the elements. Other celestial cycles and their relationship to the world of the ancients would also have been apparent. For example, the 29.5-day cycle of the moon is remarkably coincident with the menstrual cycle of women.

To help remember the time span between these recurring events would require some sort of counting system. Nothing elaborate or particularly complicated: Stones placed on a pile, daubs of ocher on a rock wall, scratched lines on a bone, or notches on a stick would have been sufficient. Perhaps cumulative counting systems appeared spontaneously among a specific group of people; more likely they developed gradually over many generations, possibly hundreds, growing out of oral traditions and appearing in different cultures at different times, depending on the needs and the intellectual readiness of a people.

To understand how—and how much—ancient societies knew about the heavens, one must first understand some basic astronomical principles. This is not to say that all preliterate societies understood the structure of the universe or the mechanics of solar system dynamics or even knew the most fundamental concepts of elementary astronomy. Ancient astronomy was essentially naked-eye observational astronomy, similar in both execution and purpose to that practiced by Boy Scouts, small-boat enthusiasts, and backyard stargazers. Much can be learned about the sky through simple—although diligent and prolonged—observation. Today we know the causes of certain observed phenomena in the sky; ancient man saw—and was concerned with—only the effects.

To begin with the basics, the earth turns on its axis once every twenty-four hours. This rotation divides the day into two periods, not always equal, of light and dark. All creatures—indeed all living things, including plants, one-celled organisms, and humans—make unconscious biological responses to this diurnal cycle. This is the first, and most obvious, astronomical phenomenon that would have been noted by ancient man.

The rotation of the earth also makes it appear as if the sun, moon, and stars rise in the east and set in the west. Because the earth is tipped slightly on its axis, the stars have another distinctive motion: Except at the equator, they seem to turn slowly around a single point, known to us as the celestial pole (see Figure 5). The farther one is from the equator, the more pronounced is this movement (see Figure 6). For modern observers, the bright star Polaris seems to stand still as the other stars wheel about it. Polaris itself has no particular significance, except that it is the brightest object near the spot where the earth's axis of rotation points into the sky. A thousand years ago another star occupied this position; and thousands of years in the future still another star will hold its place. The combined forces of lunar and solar gravity actually cause the celestial poles and equator to drift slowly in a circle over a period of twenty-six thousand years. This drift is called precession and it causes different stars to become the "north star" every few thousand years; for example, in A.D. 7000, the pole star will be Alpha Cephei. But no matter which star marked the celestial pole, it served as a beacon indicating the direction north. (Obviously, a similar effect can be observed above the southern horizon; however, for simplicity—and the fact that most civilizations developed first in the north—this discussion will concentrate on the northern hemisphere.)

Just as the earth's axis of rotation can be extended into space to form the celestial poles, the earth's equator can be projected onto the sky to form an imaginary line that parallels the daily east–west route of the stars. Modern astronomers divide this equatorial girdle—as well as all circles—into 360 degrees. This practice stems

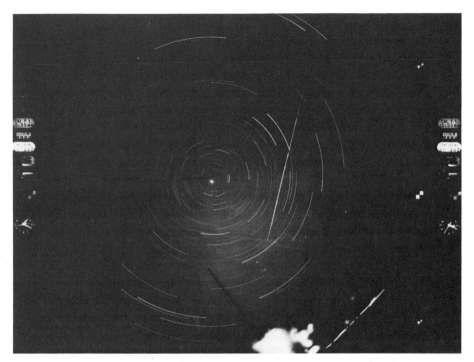

Figure 5. *The stars trace circular paths around the north celestial pole in this time exposure taken with a wide-angle meteor-patrol camera. A bright fireball appears as a streak across the image.* Photo courtesy Smithsonian Astrophysical Observatory

from Babylonian times, but may have come from a more ancient and inaccurate estimate of a year divided into 360 days. The subdivision of one degree of arc into 60 minutes and one minute of arc into 60 arc seconds (as well as that of dividing an hour into 60 minutes) comes from the easy division of 360 and 60 by a variety of integers. These numbers, still used for time, geometry, and astronomy, are the only vestiges of the ancient and somewhat awkward sexagesimal system, long since replaced for most purposes by the decimal system.

For all observers, no matter where they are on the globe, the point directly overhead is the zenith. It is also the highest apparent point reached by a star as it crosses the nighttime sky. If another imaginary line is drawn north and south from horizon to

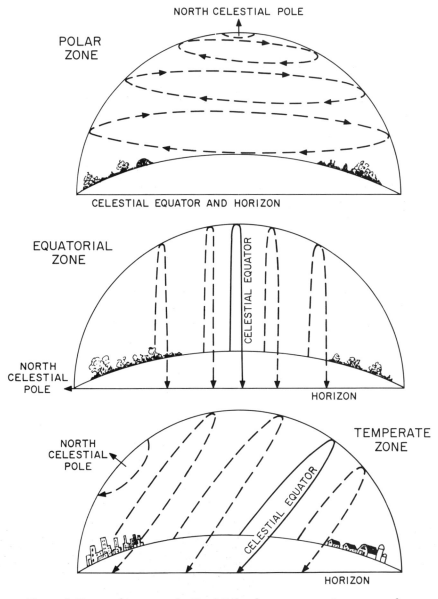

Figure 6. *For an observer at the North Pole, the stars never rise or set; rather each night the same constellations trace circular paths above the horizon. By contrast, at the equator, an observer sees the stars rise straight above the east, travel almost directly across the sky dome, and set on the exact opposite horizon, with no variation in their paths throughout the year. For observers in the north temperate zone between the pole and the equator, the stars trace*

22

horizon, through both celestial poles and the zenith, it will form the meridian, or longitude, of the observer. Again, it is uncertain if ancient peoples actually designated these imaginary lines as such, but they certainly must have known when the sun or a group of stars rose in the east, climbed the sky dome, reached their zenith, or culmination, and then, crossing the meridian, started down the other side. Observation of this simple movement would have provided a valuable way of marking time at night (see Figure 7).

Another relationship between rising stars and the sun was certainly observed by ancient peoples. Any particular star rises and sets slightly earlier each night. The first day of the year when a star can be seen just before dawn is known as its heliacal rising—that is, when it is close to the sun. One day earlier, and the star will rise after the glow of sunrise has made the sky too bright for it to be seen. On following days, the star will rise sufficiently earlier to be seen several minutes before the sun in a dark sky. (Conversely, the last night on which a star can be seen just after sunset is known as the heliacal setting.)

The heliacal risings of particular stars or star groups, because they occur at the same time each year, could mark special and important dates for a society. The Egyptians, for example, watched for the heliacal rising of Sirius as the signal of the beginning of the annual Nile flood. The Papago Indians of Arizona still traditionally divide their year by positions of the Pleiades: summer heliacal rising, time to plant; zenith at dawn, end of planting; past zenith, harvest begins; one quarter down from zenith, deer-hunting time; heliacal setting, harvest feast.

The final and perhaps most important interrelationship between sun and stars influencing ancient astronomy is caused by the tip of the earth's axis in respect to the sun. The earth, in simplest terms, is spinning like a top; and like a top it does not remain perfectly upright. The interplay of gravity and angular momentum causes

more slanting paths, rising in the east, moving slightly south to culmination, then moving slightly north again to set in the west. Moreover, the rising and setting points shift slightly each day, responding to the changing position of the earth in respect to the sun. Thus, different stars and constellations appear above the horizon at different times of the year. Illustration by John Hamwey, Smithsonian Astrophysical Observatory

23

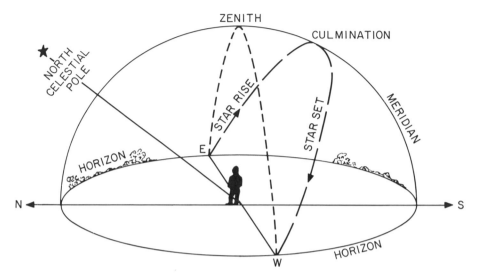

Figure 7. *For the ground-based observer, the sky dome can be adequately described by a few simple coordinates. The spot where the stars seem to stand still, or make small circular motions, is the north celestial pole. The axis of the earth points to this spot in space. The point directly above the observer's head is the zenith. A line running from the northern to the southern horizon and passing through the pole and the zenith is the meridian. The moon, sun, and stars rise in the east, reach culmination when they cross the meridian, and travel down the dome to set in the west.* Illustration by John Hamwey, Smithsonian Astrophysical Observatory

the earth to spin at a slight angle—approximately 23.5 degrees off the vertical (see Figure 8).

This tipping causes several observable effects. First, by watching the sun and stars at dawn and dusk, one can see that each day the sun shifts eastward about one degree (approximately twice the apparent diameter of the moon as seen by the naked eye) relative to the stars. This makes the sun appear to move through different clusters of stars at different times of the year. However, year after year the sun follows the same path through the same sequence of star groups. This path is called the ecliptic.

In ancient times, the star groups through which the sun traveled during the year were organized into patterns, or constellations, and together they became known as the zodiac, a Greek term

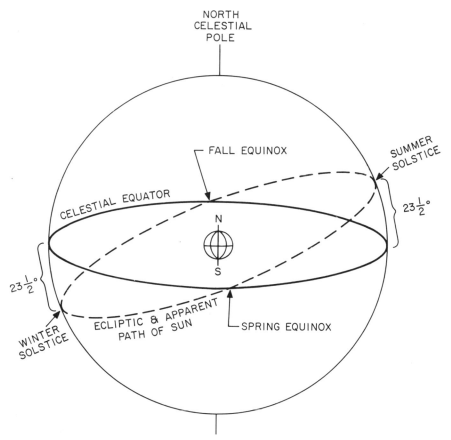

NORTH
CELESTIAL
POLE

FALL EQUINOX

SUMMER
SOLSTICE

CELESTIAL EQUATOR

$23\frac{1}{2}°$

N

S

$23\frac{1}{2}°$

WINTER
SOLSTICE

ECLIPTIC & APPARENT
PATH OF SUN

SPRING EQUINOX

Figure 8. *The earth is tipped on its axis 23.5 degrees in respect to the plane of its orbit around the sun. This orbit can also be described as the plane of the ecliptic, and twice each year it crosses the celestial equator.* Illustration by John Hamwey, Smithsonian Astrophysical Observatory

meaning "circle of animals." * Of course, different cultures saw different patterns in the sky, with each image derived from a culture's own experiences and mythologies. Many familiar star groups of the Western world may have originated in the Near East as

---

*The formal naming of the constellations, particularly those in the zodiac, once thought to be a Greek innovation, may have come two thousand years earlier. Rather than a casual, unstructured naming of stars for legendary heroes and heroines or familiar objects and animals by lonely sailors and shepherds, the zodiac design may have been a carefully conceived plan to identify and remember the

early as 3300 B.C. Artifacts from Mesopotamia show Leo chasing Taurus from the sky, an apparent reference to the rising and setting of the two constellations in that epoch.

Even if ancient man did not note the yearly movement of the sun along the ecliptic and through the zodiac stars, he must have been aware of the sun's changing position in the daytime sky due to the earth's tilt. As the earth revolves around the sun, its north pole is tilted first toward and then away from the center of its orbit. The changing position of the pole causes the seasonal changes; for instance, when the north pole points toward the sun, the northern hemisphere experiences summer and the southern hemisphere winter. The difference in seasonal warmth or cooling is not due, as many people mistakenly believe, to the northern hemisphere's being closer to the sun in summer or farther away in winter; actually the slight difference in distance is insignificant. More important is the angle at which the sun's rays strike the surface of the earth; in summer they are more direct, while in winter they are more slanting and thus lose considerable energy by absorption into the atmosphere (see Figure 9).

The mechanics of the tilting effect may not be immediately obvious to the ground-based observer; however, the changing position of the sun can easily be seen and recorded. Even the most unobservant person is aware that days are shorter and shadows longer in winter, both effects of the sun's lower angle in the southern sky. For ancient astronomers, the changing angle of the earth in relation to the sun and hence the seasonal changes could be detected by observing the yearly movement of the sun north and south of their particular location. For those peoples living at the high latitudes, for example northern Europe, this yearly movement is quite pronounced and striking. The careful watcher of sunrises and

celestial pole, equator, and rough sky coordinates. Astronomical detective work by British scientist Michael Overden shows that the current zodiacal constellations correspond to positions appropriate for the celestial equator around 3000 B.C. Calculating the shift of the celestial pole due to precession, Overden found that the ancient constellations show a symmetry around Alpha Draconis, now about 25 degrees from the north position, but a potential pole star around 2600 B.C. Overden contends that the constellations were established by some seafaring people who needed guides for navigation. The most likely candidate is the Minoans.

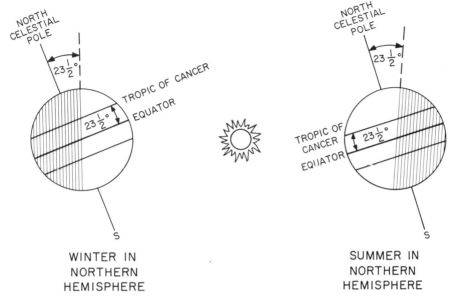

Figure 9. *The 23.5-degree tilt of the earth on its axis is the principal cause of the seasons. As it orbits the sun, the earth's axis remains pointed in the same direction; thus, summer comes to the Northern Hemisphere when the sun's rays fall more directly on that half of the globe above the equator, and winter comes when the rays are more slanting.* Illustration by John Hamwey, Smithsonian Astrophysical Observatory

sunsets would soon see the sun's changing positions and recognize the repetition of its motion over several years.

In today's air-conditioned and central-heated world, the changing seasons are only superficial reminders of passing time rather than guides to survival. However, prehistoric hunters (and, until this century, most modern agricultural societies) were slaves to their environments. Seasonal changes signaled the need for vital actions—planting, hunting, and storing. (Ironically, the energy crisis may strip modern man of many of the advantages he has gained. Today's humans could find they have a new dependence on natural cycles—and a new relationship with their ancient ancestors.) For most ancient societies, four dates in the year, each marked by rising or setting positions of the sun, became vitally important in determining seasonal changes (see Figure 10). (Again, the dates given here are all for northern-hemisphere sites.)

27

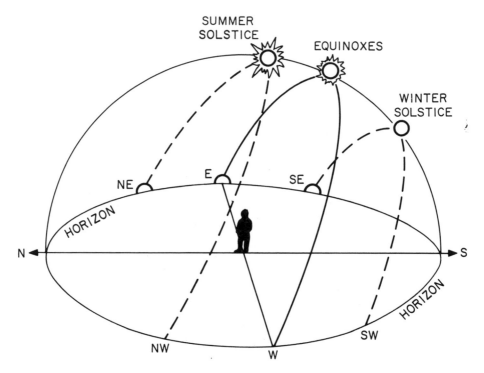

Figure 10. *In the north temperate zone, say at the latitude of Stonehenge, an observer can easily realize that the shifting positions of the sunrise and sunset points are related to the changing seasons. At the summer solstice, the sun rises at its extreme northerly point on the eastern horizon, passes high overhead at noon, and sets at its extreme northerly position to the west. At the winter solstice, the sun rises far to the southeast, remains low in the southern sky at noon, and sets at an extreme southerly position on the western horizon. At the spring and autumn equinoxes, the sun rises and sets almost due east and west.* Illustrations by John Hamwey, Smithsonian Astrophysical Observatory

The first was the spring equinox, usually occurring on or around March 21. On this date the sun crosses the celestial equator and moves north in the sky. The sun on this day appears to rise due east and set due west. Today we call this the first day of spring; for ancient man, it was the day when the sun "returned" to the sky over his lands and thus marked the return of warmth and life. On this day, too, the hours of night and day are approximately equal, and agricultural societies knew that the land should be prepared for planting. This was a time of renewal and rebirth. It is no coincidence that the Christian celebration of Easter falls near this date.

The second day, and the most important in many prehistoric calendars (especially those of northern peoples), was the summer solstice, on or around June 21. On this date—now called the first day of summer in the United States, and midsummer in Europe—the sun reaches its farthest point north of the celestial equator. The rising and setting points on the horizon are at their greatest northern extremes, and for a few days the sun seems to stand still—that is, rising at almost the same, most northerly, point on the horizon. Since this day has the most hours of daylight, the summer solstice has been traditionally celebrated by great orgiastic revels, marked by bonfires, all-night drinking and dancing, and unbridled lovemaking. Somewhat modernized and watered-down versions of this celebration are still observed today in the Scandinavian countries.

Philip Morrison has noted that almost every culture has some sort of midsummer's night festival, and if the United States did not have its Fourth of July, it certainly would have invented some other holiday for that time of year.

The next important date was the autumnal equinox, on or around September 21, when the sun crosses the celestial equator moving south. In many ancient societies, this date marked the time for harvest and, of crucial importance, when to prepare for the cold weather and lean days ahead.

The final date in most ancient calendars was on or around December 21, now noted as the first day of winter (or midwinter, in Britain). This is when the sun is farthest south of the celestial equator, so its sunrise and sunset positions are the opposite extremes of the midsummer rising and setting. The hours of nighttime are longest, and for the northern latitudes a long, hard period of poor weather lies ahead. Yet an ancient skywatcher could take hope in one fact: The sun now would not go any farther south, and it must now return north, beginning the seasonal cycle once more.

Careful and repeated observation of the sun at equinoxes and solstices would have eventually established a sense of the four seasons. More detailed descriptions of the year might be achieved by a simple counting of days between solstices. However, since there is some uncertainty around the solstice time as to what is the exact date, and because the earth's true rate of rotation is 365 and *one-quarter* days, calendars based on such simple counts are soon out of phase with the seasons.

A calendar based on lunar cycles also becomes unsynchronized even more rapidly, because the moon's 29.5-day cycle produces either a 12-month year of 354 days or a 13-month year of 383.5 days. Still, the moon is a handy—and very obvious—calendar, providing a "medium" period between the "short" day-count and the "long" year-count. For the earliest and most primitive societies, such as that of Cro-Magnon man, the discrepancy between moon year and sun year probably made little difference, since their moon counting may have been only the crudest kind of timekeeping. For more advanced cultures, such as the Egyptian, the desire for a more accurate calendar led to observation of the heliacal rising of stars to measure the year's length. For the Mayas, the desire for accuracy became a societal near obsession and produced elaborate interlocking calendars involving sun and moon cycles as well as the observation of stars and planets.

The moon, orbiting the earth as the earth orbits the sun, causes special problems for observers because its motion is so different from that of the sun and stars. Because the moon's plane of orbit is also tipped relative to earth, just as the earth is tipped relative to the sun, the paths of all three bodies periodically cross each other (see Figure 11). When the earth comes between sun and moon, a lunar eclipse may occur. When the moon comes between earth and sun, a solar eclipse may occur (see Figure 12). (Through a quirk of nature and perspective, from the distance of the earth the diameter of the moon appears identical to—or slightly larger than—that of the sun, although the latter body is actually millions of times larger.) Because the moon's motion and the problem of predicting eclipses is so central to any explanation of Stonehenge, detailed discussion will be saved for the next chapter. However, it is worth noting that the complicated nature of the moon's behavior apparently so intrigued the ancient societies of the British Isles that they constructed devices to observe and record it systematically.

30

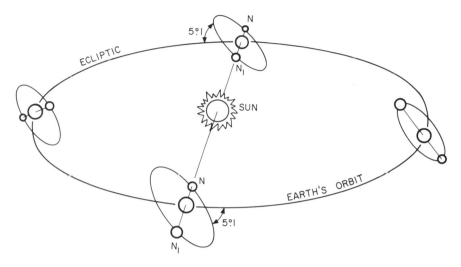

Figure 11. *The moon's orbit is tipped to the earth in respect to the plane of the ecliptic by approximately 5 degrees (greatly exaggerated in this diagram). As the earth orbits the sun, there are only two periods during the year when earth, moon, and sun are properly aligned to produce eclipses. The points where the moon's orbit crosses the ecliptic are known as the nodes (N and N$_1$); when the line N–N$_1$ points to the sun, an eclipse can occur.* Illustration by John Hamwey, Smithsonian Astrophysical Observatory

To ancient observers, the other celestial objects with erratic motions were the planets. Five planets are visible with the naked eye: Mercury, Venus, Mars, Saturn, and Jupiter. The days of the week still take their names from the seven naked-eye objects of antiquity: *Sun*day and *Mon*day from the sun and moon, *Tues*day from Tiu, the Norse name for Mars (in French, it's even more obvious: *mar*di); *Wednes*day from Woden, Norse for Mercury (in Spanish, *miér*coles); *Thurs*day, for Thor, or Jupiter; *Fri*day for Freya, or Venus (in Italian, *ven*erdi); and *Satur*day, from Saturn.

The earliest skywatchers must have quickly discovered that these five starlike objects in the sky were not "fixed" as were the other stars. They moved independently and somewhat unpredictably through the other stars as wanderers. However, they never roamed far from the ecliptic, the path that the sun followed through the stars. In fact, since all the visible planets have roughly the same orbital planes as the earth, they appear to travel within

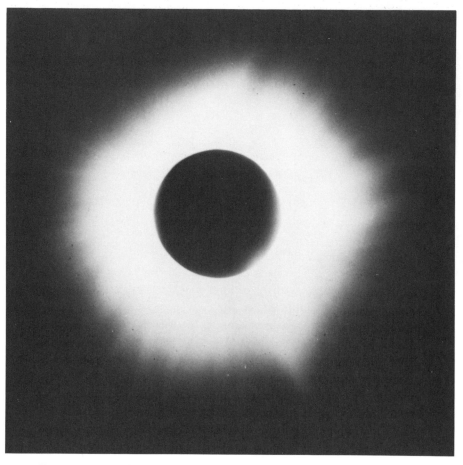

Figure 12. *A total solar eclipse showing the sun's gaseous outer atmosphere, or corona.* Photo courtesy Harvard College Observatory

the broad band of the zodiac, never straying more than 11 degrees from its center.

Ancient astronomers also saw other unpredictable and sometimes spectacular denizens of the heavens: comets, meteors, and novae. Comets are small conglomerates of dust, gas, and water vapor frozen into what astronomer Fred Whipple has described as "dirty snowballs." These bodies, apparently formed in the early phases of solar-system history, follow great elliptical paths around the sun that may take them out beyond the most distant planets.

Figure 13. *Comet Ikeya-Seki photographed in 1965 when its tail stretched several million miles into space.* Photo courtesy Smithsonian Astrophysical Observatory

When they periodically return to the vicinity of the sun (some make only one close approach in tens of thousands of years; others return as often as every three or four years), solar radiation causes the frozen gases to sublimate so the cometary nucleus (sometimes no more than a few yards in diameter) brightens and a tail of gas and dust millions of miles long streams out in a direction away from the sun (see Figure 13). Comet symbols abound in both ancient and contemporary preliterate art, including Egyptian paintings, rock carvings of the American Southwest, and the straw constructions of the Tarascan Indians who still live around Mexico's Lake Pátzcuaro. Still, it is unlikely that any ancient astronomer understood the principles of the comet phenomenon, or ever determined the periodicity of any comet.

Novae are exploding stars. In the last stage of stellar evolution, a star's fuel supply may become so depleted that the delicate equilibrium between internal pressure and gravity can no longer be maintained and the star explodes in one last gasp of outrushing gas and dust. Millions of tons of matter are spewed into space in seconds. Such a star explosion, especially if it is a supernova in our own galaxy, can be dramatic, often marked by an increase in brightness of tens of magnitudes. An insignificant or even previously invisible object may suddenly become a brilliant point in the sky, visible even in daylight. The brightening may persist for several days or even weeks, until the gas and dust have been sufficiently diffused into space and the shrunken dark core of the star disappears again. Current theory suggests that at least one super-nova should be seen in our Milky Way galaxy every one hundred years. To a diligent observer of the sky familiar with star patterns, both novae and comets should appear as "new" (from which derives the name *nova*) or "guest" stars. Evidence that ancient astronomers observed novae exists in the records of court astronomers of medieval China. A less conclusive record of rock paintings and carvings in the American Southwest suggests that the appearance of certain guest stars was also noted by American Indian tribes.

Meteors, too, form a special class of celestial objects for which the evidence of ancient observation is almost totally presumptive. Millions of tons of cosmic debris—the leftover bits and pieces of the solar system, including minute particles of dust, rocky chunks of asteroids, and icy droplets of comets—fall on earth each year. While still in space and orbiting the sun, these objects are known as meteoroids. When their paths intersect the orbit of earth and they are drawn into its atmosphere, the friction of their passage heats the atmosphere around them to produce light, an effect known as a meteor. The common "shooting stars" seen by every backyard barbecuer certainly must have been seen by ancient sky-watchers.

Sometimes the entering body is so large it creates a pyrotechnic display complete with multicolored lights as bright as the full moon, sparks, a visible smoke trail, and even sonic booms. Usually even the largest of these fireballs (or bolides, as the exploding variety is called) disintegrates into a rain of dust and small particles before reaching earth. Occasionally, however, a sizable object survives the fall and lands intact on the earth's surface. A rare few may be large enough to strike with enough velocity to produce a crater. Some craters, such as Winslow Crater in northern Arizona, have been formed since *Homo sapiens* inhabited the earth, and it is reasonable to assume such events were witnessed. But the record is unclear.

This is due partly to the transient, infrequent, and highly localized nature of meteoritic events. Even the brightest fireball occurs so low in the atmosphere and so close to the ground observer that the curvature of the earth limits the area from which the meteor can be seen. Usually the diameter of this area of visibility is no more than a few hundred miles. Since truly bright meteors are also once-in-a-lifetime events, there might be no repetitions over several generations to reinforce oral traditions. Even if recorded, the single record of a unique event by an isolated group of people would stand a good chance of having been erased by either successive civilizations or climate changes.

However, some indirect evidence suggests preliterate man did make a connection between meteorites and their celestial origin. There are two broad classes of meteorites: the "stones" and the "irons." (Actually, both types contain inordinate amounts of iron; this is one of the characteristics that distinguish them from common field rocks. But the iron meteorite is almost purely that element.) Examples of tools and particularly of ceremonial blades made from high-grade iron found among Stone Age and Bronze Age artifacts suggest that iron meteorites were recovered and recognized as unusual. Since many of these objects have been found in sacred or ritualistic contexts, ancient man may have considered them as otherworldly and possessing supernatural powers. The Kaaba, the sacred "black stone" of Mecca, actually may be a large iron meteorite.

The question that stalks archaeoastronomy is not if early man was aware of these various phenomena in the sky, for even simple animals respond to certain celestial events, but rather when he first became aware that some had a cyclic nature, and began to make practical use of his knowledge. When did he discover that celestial events could be time-factoring devices? According to some researchers, this cognitive ability may have emerged as early as thirty thousand years ago.

Sometime, perhaps more than 2 million years ago, somewhere in East Africa, an apelike creature called *Homo habilis* took a giant step toward human evolution by using chunks of chipped rock and pieces of sharpened bone as crude tools.

Homo habilis, "the handy man," still a rather rough approximation of the creature he might eventually become, was replaced on the evolutionary path to the future by *Homo erectus*, who appeared just a little less than a million years ago. This species of advanced but still simple tool-makers gave way in turn to the brutish-looking but apparently intelligent *Homo neanderthalensis* some seventy-five thousand years ago. And it is with Neanderthal man that we see the first evidence of cultural and ritual behavior as opposed to strictly utilitarian activity. Moreover, Neanderthal man may also have developed a real awareness of the vital link between human life and natural forces.

At Shanidar Cave in the Zagros Mountains of Iraq the remains of a sixty-thousand-year-old Neanderthal burial site have been found. Pollen grains discovered amid the rubble of the cave floor suggest that the body, perhaps that of a chieftain or elder, was surrounded by several different kinds of wildflowers. The distribution of the pollen grains suggests the flowers were deliberately and carefully laid out, not simply thrown haphazardly over the body. More important, several of the plants have been identified as the same species used as herbal medicines for centuries by primitive societies. "That the Shanidar people were aware of at least some of the medicinal properties of the flowers is not unlikely," writes Richard Leakey in his book *Origins*. "As they became more and more independent of the demands of the environment, early members of the human family must also have had an intimate knowledge of what nature offers. This, indeed, is the key to success among hunting and gathering communities."

Neanderthal man, of course, would lose the evolutionary battle to an even more successful species, Cro-Magnon man. The earliest known representatives of this species, otherwise known as *Homo sapiens*, probably originated in southwest Asia and gradually made their way westward to northern Europe. Some forty thousand years ago, they began to displace Neanderthal man.

The Cro-Magnons were tall, well built, smooth skinned, and handsome; in fact, they looked very much like modern human beings. Some physical anthropologists think a Cro-Magnon man dressed in modern clothes could walk the streets of Western Europe today without attracting any undue attention.

Some forty thousand to ten thousand years ago, just before the last Ice Age, Cro-Magnon people lived as hunters and gatherers in the northern temperate zone of Europe and Asia, thriving on the abundant game pushed southward by the advancing glaciers. Despite their undeserved caveman image, the Cro-Magnons developed a sophisticated and complex society and produced artwork of exquisite beauty. "In these artworks, and in the stone tools and other artifacts that have been found with them," writes Boyce Rensberger in a *New York Times Magazine* article, "lie all we are likely to know of the roots of our own culture; for the Cro-Magnon were us. . . . Their brains were every bit as large as ours and presumably they were capable of doing everything we can do. Whatever we learn about the Cro-Magnon, we learn about ourselves for it is to their way of life that we are biologically adapted."

The best-known records of the Cro-Magnon people are the cave paintings of southwestern Europe, most notably those at Lascaux in France and Altamira in Spain. When these artifacts were first discovered and described nearly a century ago, most archaeologists refused to believe the paintings were the work of primitive man.

The traditional view of Cro-Magnon man was that of a loutish brute with little intellectual capacity. Yet the cave art stunned observers with its technical accuracy and aesthetic grace. Great herds of bison and elk galloped across rock walls at full speed, their legs a blur of motion. Wounded stags cried out in pain. Other animals scratched at cuts or flicked at insects on their backs. The line of a deer's back flowed with all the effortless grace and economy of a Chinese scroll painting. Although animal forms were rendered with realistic fidelity, the artists also used abstractions for human forms and other, still undeciphered ideas. They also mastered complex graphic techniques such as perspective, motion, multiple figures, and dimension. Most of these artistic effects would not appear again for another twenty thousand years. No wonder, then, that many experts considered the cave paintings to be clever modern frauds.

Although artists and historians have studied the cave paintings and other artworks, including incised antlers and bones and small stone carvings, for nearly half a century, archaeologists and anthropologists have only recently come to appreciate the extraordinary technical skills and organizational abilities that made these creations possible.

The Cro-Magnon painters used at least a half dozen different pigments—reds, browns, yellows, and black—all powders of natural earth mixed with binders such as animal fat, egg white, blood, vegetable juices, and a primitive glue made by boiling down fish scraps to a soupy paste. Moreover, the artists apparently used the different binders for different surfaces, choosing carefully their materials to match the conditions of the rock—dry or wet, hard or porous.

For brushes, Cro-Magnon artists used animal hairs or the frayed ends of twigs. The technique of applying paint varied, again for different purposes and different surfaces. Sometimes paint was daubed on with fur pads; at other times, it was smeared on thickly with a flat blade, just as modern painters use palette knives. Paint might be smeared with fingertips or lightly applied with a semidry brush to produce shading effects. Some pigments were allowed to dry and harden so the color could be used like crayons. Flat rocks served as palettes, old shells as paint jars. Rough sketches in charcoal were done on flat rocks or scratched on sheets of slate. To illuminate their efforts—and most of the work was done deep within the interiors of caves—the artists used crude oil lamps, usually made of hollowed-out rock with a wick of twisted reeds floating in a pool of animal fat.

"Quite aside from the fact that the very idea of making a painting required a high order of conceptual ability," writes Rensberger in his *Times* article, "it must have taken a great deal of planning and skill to gather and mix the minerals and binders for the paints, prepare the tools and lamps (oil from animals, for example, is mainly harvested in the late autumn when animals have fattened up for winter), and make preliminary sketches. And all for what? This is the great enigma. . . ."

The purpose of this art, most of it hidden in nearly inaccessible caverns, stirs heated debate in the archaeological community. The general view is that the paintings had some religious or ceremonial purpose linked in a general way to the hunting and killing of the animals depicted on the walls. Similar artistic obsessions are still found among preliterate peoples today for whom a certain animal or group of animals is central to their economy. Patricia Vinnecomb has found the Kung Bushmen of the Kalahari Desert endow many animals, and particularly the eland, with mystical powers. These animals are represented over and over again in Bushman art.

The presence of art in deep caves also suggests that these caverns were secret and sacred places—the catacombs of the Ice Age, perhaps—to which the hunter-priests returned repeatedly for special ceremonies. New evidence based on improved techniques of lighting and photography both supports the theory of ritual centers and points toward the emergence of man's cognitive abilities. Indeed, understanding *how* the paintings were produced may prove more important than *why* they were created.

Through an analysis of ultraviolet and infrared photographs of the cave paintings, Alexander Marshack of Harvard University's Peabody Museum has found that much of the art was redone several times over, with underdrawings and additions by several different artists at different times. Marshack feels the retouching and repeated stroking, often in quite different styles, implies that the paintings had a periodic use in some recurring ritual.

It is this possible periodic use of the ceremonial centers that most interests Marshack, for it suggests the artists must have had some sense of time, even if it was only the passing of the seasons. This sense would have not only brought them back to the caves at predetermined intervals, but could have led to the development of other cognitive skills, such as counting and record-keeping. For more than fifteen years Marshack has been searching for examples of how prehistoric peoples measured time. His findings, although still controversial, are extraordinary, for they suggest that systems of time notation may have existed at least twenty-five thousand years before the cuneiforms of Mesopotamia or the hieroglyphs of Egypt. Oddly, Marshack, by training and background a journalist rather than an archaeologist, came to this conclusion by chance.

In the early 1960s, while working on a book about the United States moon mission, he happened to see an article in *Scientific American* describing an eighty-five-hundred-year-old engraved bone tool from Ishango, in the Congo. The author of the article interpreted the markings on the bone as some sort of arithmetical game. Marshack, perhaps because of his immersion in the lunar project, saw the marks as a time count, with the 168 lines representing the lunar phases over a six-month period.

One question that has tantalized historians—and journalists—is how, and why, civilization spontaneously burst forth in the valleys of the Nile and Tigris-Euphrates rivers. How could human beings suddenly develop the mathematical and engineering skills to make these civilizations possible? Had the human race received help from outer space, as suggested by some writers? Had the species been boosted to a new intellectual state due to mutation caused by cosmic rays, as suggested by others? Or had there been a slow, steady development of human cognitive skills over thousands of years, with the record of this evolution lost, overlooked, or misinterpreted?

Marshack suspected the latter, with the Ishango carving a case in point. He felt that "civilization may have been built as much on time-factoring skills as on the hand-made tools that we find in layers of soil." With financial support from the National Science Foundation and the Wenner-Gren Foundation, and moral support from Harvard, Marshack abandoned his moon book and traveled to Europe to look for other records of notational markings on the prehistoric artifacts in museum collections. He carried with him a small pocket microscope to examine the faint scratches found on many carved bones and previously dismissed as no more than decorative doodles. As he wrote in his *The Roots of Civilization*, Marshack believed that "with the microscope one could determine how many tools or points or grips or styles of strokes were used to make a sequence of marks. This could help in deciding if a composition was merely decorative or notational. For, if it was made at one or two sittings with one or two tools, with one concept and one rhythm, it could be decorative. If it was made by many points over a long time, it was probably not decorative."

Throughout the collections of Europe, Marshack found dozens of bones and antler chunks dating from the Cro-Magnon period that were inscribed with grooves and furrows, many still bearing traces of red ocher as if they once were filled in to highlight and emphasize the lines. On many of the bones, the shapes and the grouping of the gouges were reminiscent of both the Ishango bone and the counting sticks found in some modern preliterate cultures. Marshack concluded that these lines were not random or capricious, but rather careful, thoughtful records of the lunar month.

In support of his somewhat enigmatic system of lines and gouges in bone, Marshack also inspected and analyzed the more representational carvings of flora and fauna found on other bone fragments. He again found that these pictures were not simply random outpourings of artistic creativity. Rather, the images of certain animals and plants seemed grouped by season, with spawning salmon, rutting elk, and budding flowers often appearing together as part of the same general scene. Marshack was the first investigator to note this seasonal grouping, a relationship that now seems so obvious it is surprising that no archaeologist had seen it earlier.

Marshack also found many flat stones carved with curved lines apparently symbolizing female forms. Some of these forms later had been "crossed out," as he described it, perhaps as records of childbirths. To Marshack, the seasonal groups of animals suggest that prehistoric man observed the natural cycles around him, and, in the childbirth stones, attempted to count and record certain events.

More pertinent to the study of archaeoastronomy is Marshack's theory on lunar notation. The moon is the most prominent astronomical body with an observable cycle. It is very likely that Cro-Magnon man observed the moon and had the brain power necessary to record its monthly phase changes. In fact, however, because of a variety of reasons, the observed lunar month—from first crescent to first crescent—sometimes may be as short as twenty-eight days or as long as thirty to thirty-three days. Marshack's interpretation is actually based on an average lunar month of his own devising that takes into consideration the possibility that bad weather could have prevented the observer from seeing the first crescent or that seasonal changes in the moon's orbit relative to earth would have made it more difficult to observe risings or settings from a specific location. When his hypothetical lunar observing record, or "template of notations" is applied to the bone scratches, there seems to be a remarkable correlation between marks and moon. If one accepts Marshack's weighted averages, the bone notations become a clear record of time-factoring activity in a primitive society.

Of course, Marshack's conclusions have not been universally accepted. His background as a journalist, screenwriter, and popularizer of science does not help win support in the scientific community. But even beyond this somewhat irrational response, many critics justifiably worry about the numbers game that is possible by juggling the many variations in the lunar cycle. The scratches on bone could be interpreted to match several proposed cycles—or theories.

Still, Marshack's general ideas are steadily winning support throughout the scientific community. Several researchers have turned to contemporary preliterate societies for clues to how a people's cognitive abilities may become quite sophisticated without the development of any technical skills. For example, Howard Morphy of the Australian National University, through the development of a close personal relationship with an aboriginal chief, found that many of the seemingly meaningless marks on bark carvings, totally incomprehensible to a nonaborigine, really were abstract symbols providing complicated information about hunting and survival techniques. Indeed, some were very explicit maps to waterholes and game areas. Yet in many cases, even other aborigines couldn't read the message, for the code was passed on orally so that only those people with "the right to knowledge" would know the meaning.

Travis Hudson, an anthropologist at the Santa Barbara Museum of Natural History, has looked at the artifacts of those Indian tribes living in California before the Conquest and has found evidence that the shaman-astronomers had "compiled a corpus of astronomical information and devised explanations for what they observed in the sky to an extent far beyond what has been previously credited to hunters-gatherers." This early appearance of astronomical knowledge and skill in preliterate cultures has been previously overlooked, Hudson claimed in a presentation to the AAAS in 1980, because of the deep-rooted conviction among many anthropologists that "if people lacked crops or flocks, they lacked a sophisticated awareness of the heavens. Expressed another way, astronomy and calendrics were the inventions of food producers." His recent research on sun-watching sites, calendar devices, and surviving sky legends suggests the opposite. The hunting-gathering tribes of pre-Spanish California, although no more technically advanced than Marshack's bone-carvers, "performed a variety of stellar, lunar, and solar observations, using the data to regulate economic, political, and ritual behavior."

Marshack's real contribution, particularly for archaeoastronomers, is the suggestion—still not proven—that the intellectual development of man and his cognitive powers, once thought to have begun only with the establishment of agricultural centers some ten thousand years ago, may actually have begun twenty thousand years earlier. While the recording of lunar-cycle scratches on a deer antler may seem a long way from erecting giant sighting stones, the steps are not really that far removed. Once the basic foundation of astronomy was laid down by the discovery of a counting system, the several thousands of generations of people that followed would have had both ample observing experience and a store of cumulative knowledge necessary to erect permanent and sophisticated observatories.

Today much of this Ice Age evidence seems very obvious; however, when Gerald Hawkins tried to suggest that such basic intelligence had been responsible for the building of Stonehenge, he touched off a debate about ancient peoples and cultures that continues today.

The Stonehenge Connection

"Time-hallow'd pile, by simple builders rear'd
Mysterious round, through distant times rever'd!
Ordained with earth's revolving orb to last!
Thou bring'st to sight the present and the past."

—John Ogilvie
The Fane of the Druids (1787)

In the fall of 1963, while I was toiling as the Information Officer of the Smithsonian Astrophysical Observatory in Cambridge, Massachusetts, my colleague John B. White told me about a staff member who was working on a new interpretation of Stonehenge. Because it was not the prime concern of the scientist, he was doing the work in his spare time and the results had languished on his desk for months, until White persuaded him to publish at least a summary. This had been done and a preprint was ready for submission to *Nature*, a major British scientific journal. In the paper Gerald Hawkins, an astronomer who held joint appointments at the Harvard and Smithsonian observatories and at Boston University, proposed that Stonehenge might have been intended as an astronomical observatory by its prehistoric builders. Hawkins's theory was based on a computer analysis of the alignments of that ancient monument's stones and the positions of the rising and setting sun and moon during the epoch estimated for Stonehenge's construction.

As a journalist I realized the great public appeal of this paper. Stonehenge, mysterious and silent, its origins and intentions lost in prehistory, had produced speculation, both romantic and scientific, for years. Now, here was a reasonable and plausible explanation for its creation put forward by a professor of astronomy with an impressive publications record and the directorship of a university observatory. Better yet, the explanation came about through the use of a massive high-speed electronic computer at a time when the public still held such devices in awe. An ancient mystery solved by space technology!

My news release reflected this theme:

> Using only one minute of borrowed computer time, Dr. Gerald Hawkins may have found one answer to the 4,000-year-old mystery of Stonehenge. With 1500 B.C. as a probable date . . . Hawkins gave an **IBM 7090** computer the task of looking for correlations between the directions defined by the lines joining various stones and holes and the directions of the rising and setting of the sun and moon at midsummer and midwinter. . . . The results show 10 definite correlations with solar directions within an accuracy of 1 degree and 14 correlations with lunar directions within 1½ degrees. The chance of such a correlation being purely coincidental is roughly one in a million. . . . Though the anthropological reasons for construction of Stonehenge are still speculative, Hawkins feels the monument could "form a reliable calendar for predicting the seasons. It also could signal the danger period for eclipses of the sun and moon."

Hawkins had found these correlations himself, using pencil, pad, slide rule, and logarithm table, and used the computer to check and recheck the hand calculations rapidly and with absolute accuracy. The release was three pages long, with a diagram, and the accompanying journal article carried tables of data supporting the academic discussion; but, in essence, this was the premise: Stonehenge was deliberately and consciously constructed as an observatory.

Although I had instinctively sensed the potential public response to any theory linking forgotten peoples by an IBM computer, I don't think I was really prepared for the magnitude of that response. The news release was picked up by major news services and broadcast media. Hawkins was besieged with requests for interviews and comments. Producers from a television network proposed a one-hour documentary special and a major publisher suggested that the research be turned into a full-length popular book. But if I had underestimated the popular reaction, I never expected the vehemence and ferocity of the opposition from certain opinion leaders in the British archaeological establishment. The most outspoken critic would be Richard J. C. Atkinson of University College, Cardiff, Wales, the leading authority on Stonehenge and its era in prehistory.

Hawkins's interpretation necessarily implied that the builders of Stonehenge had cognitive skills sufficient to understand and appreciate celestial cycles, and organizational skills sophisticated enough to plan a major construction project laid out in a precise and predetermined fashion. Assuming that Stonehenge was built around 2000 B.C., his suggestion flew in the face of all other archaeological evidence, which said the prehistoric Britons of this era were screaming savages whose material culture had not progressed beyond rough earthen pottery and rude stone tools. If correct, Hawkins's theory required a reexamination of history—not to mention the careers of several major archaeologists! With one minute of computer calculations, Hawkins had refuted nearly a century of archaeological dogma. He had, as one writer noted, "overturned the archaeological applecart" and begun a revolution that would lead to a complete rethinking of the intellectual development of humankind.

Of course, public interest—and scientific controversy—over Stonehenge and its purposes did not begin with Hawkins. Stonehenge had long been part of the popular and scientific imagination. Several scientific reputations had been made—and ruined—by speculations about Stonehenge long before Hawkins fed his data into the computer. Scores of dramas, novels, and poems used Stonehenge as an appropriately exotic setting. Religious cults had developed and thrived beneath the hanging stones. Indeed, even the possible connection between Stonehenge and astronomy, particularly its relationship to the midsummer sunrise, had been suspected for generations. What was it about this circle of rough-hewn rock that people found so intriguing?

To the modern visitor, especially one who has been immersed in Stonehenge fact and fancy, the stone ring can be a disappointment. Conditioned in part by contrived photos, purple-prose descriptions, and one's own imagination, the preconceived image of Stonehenge tends to grow in the mind to several times life-size. In fact, surrounded by the green, open airiness of Salisbury Plain, Stonehenge looks positively minuscule and somewhat drab. Cows grazing in nearby fields create a benign bucolic atmosphere rather than any feeling of malevolent brooding or mystery. Worse yet is the sad discovery that little of the original ring still stands. Most stones lie, as the archaeologists say, recumbent, or have disappeared over the centuries. Somehow one expects to find the circle intact. Yet standing inside the uneven circle, especially at dawn when the rays of the rising sun poke through the gaps between the standing stones, one can understand the wonder and awe of earlier visitors. (Alas, within the past few years most visitors have been prevented from experiencing this thrill. A fence now surrounds the stones and prohibits entry to the inner circle: a long overdue precaution designed to save the remaining stones from being worn and chipped away by the ever-growing number of tourists.)

Only extended and repeated visits reveal the true size and complexity of the site. Stonehenge is in fact a series of concentric circles, encompassing much more than the main ring of large gray stones that first catch the eye (see Figure 14). The first and largest circle is a grassy ditch a few feet deep and approximately one hundred meters in diameter. Immediately inside the ditch is a grassy rampart, or earthen ring, now about two meters high and presumably constructed from the dirt taken out of the ditch. Just inside the earthwork is a circle of holes, known as the Aubrey holes for their seventeenth-century discoverer, John Aubrey.

Two other rings of holes—the outer known as the Y holes; the inner, the Z holes—are about twenty meters from the earthen

Figure 14. *A section of the sarsen circle with upright stones and lintels in place. The man at the center of the photo is standing beside one of the blue stones of the inner circle.* Photo courtesy C. A. Federer

ring. The two rings are aligned so that the holes seem to radiate out from the standing stones at the center of the complex like spokes of a wheel. Their original purpose is unknown.

Next is the sarsen circle of stones, the most familiar feature of Stonehenge. At one time, this circle consisted of thirty upright stones, each weighing nearly twenty-five tons and hewn to roughly rectangular shape. On top of each pair of stones rested a third lintel stone of similar shape and size, thus creating a series of crude arches, with the lintels carved slightly to fit the gentle curve of a circle thirty meters in diameter. The lintels were held in place by a double-jointing system, using a mortise and tenon to grip the horizontal lintel to its vertical supports and tenon and groove to link the ends of the lintels in a continuous architrave.

Over half of the sarsen stones (their name is said to derive from *Saracen,* which came to mean "foreign") have been quarried away by anonymous local farmers and builders, and experts suspect only three of the remaining sarsens are still in their original places. At the turn of the century, many stones were leaning precariously due to weathering and ground slippage. With assistance from archaeologists, sixteen uprights are now back in place and three of the fallen lintels have been restored.

The sarsen stones ranged in length from fourteen to eighteen feet, with each sunk in the earth to an appropriate depth so that each stood about thirteen feet above ground level. When the lintels were placed on top, the ring was approximately 15.5 feet high. These stones are Tertiary sandstone found around the Salisbury Plain area. Locally, the stones are described as "Grey Wethers" because of their resemblance to flocks of grazing sheep.

Within this circle of stones once stood yet another ring, about twenty-three meters in diameter—the bluestones. The bluestones are not native to the Wessex area and were long a source of some mystery—until their origin in the Prescelly Mountains of Wales was identified. The bluestones, a flinty dolomite, were considerably smaller and more slender than the sarsens and as many as sixty-one (or as few as fifty-nine—experts differ on this point) may have once formed the ring. Within the double circle of sarsens and bluestones stood five huge trilithons arranged in a horseshoe pattern open to the north-northeast. As their name implies, the trilithons were freestanding arches formed by three stones, two upright and a third held horizontally by mortise-and-tenon joints (see Figure 15).

The trilithons, of a sandstone similar to the sarsens', are impressive by any standards, with the uprights averaging fifty tons apiece and standing twenty to twenty-five feet high. The largest of these stones is also the largest example of prehistoric worked stone found

Figure 15. *One of the trilithons at Stonehenge.* Photo courtesy C. A. Federer

in Britain. Although the trilithons have withstood the ravages of time slightly better than the outer sarsen ring—the two southeastern trilithon arches apparently never fell—the largest arch collapsed centuries ago and one of the stones broke in two. The surviving upright was reset in modern times (see Figure 16).

Within the trilithon horseshoe was still another, smaller horseshoe, made of nineteen slender bluestones. Only twelve of these stones still remain. Inside both horseshoes and on the axis of the complex lies a large flat slab, popularly called the altar stone, al-

Figure 16. *Stonehenge, before tourists were banned from the inner circle, showing a trilithon* (left) *and a portion of the sarsen circle.* Photo courtesy C. A. Federer

though any connection with rituals is pure speculation. This stone, too, apparently was brought to Stonehenge from a distant location, most likely a site near Milford Haven in Wales.

Several other features at Stonehenge have particular significance when its theoretical use as an astronomical observatory is considered. Four station stones once stood at the inner perimeter of the earthen ring. Lines connecting the four positions form a near-perfect rectangle whose diagonals intersect at the center of the sarsen circle near the altar stone. Only two of these stones (stones 91 and 93 on official charts) still remain. The two other station stones (92 and 94) were centered on low mounds inside the earthen rings, mounds that also covered several Aubrey holes; their positions are known only from holes.

The earthen ring has a gap in the north-northeast which is known as the causeway entrance. The causeway itself is a flat earthen ramp that spans the outer ditch and leads from the Aubrey circle to a long "avenue" extending north-northeast from

Figure 17. *Stonehenge at dusk with the heel stone in the right foreground.*
Photo courtesy British Tourist Authority

Stonehenge for a distance of some 120 meters. Within the causeway entrance, and lying between the first and last of the Aubrey
holes, is a large recumbent sarsen known as the slaughter stone,
although again there is no real evidence to support such a purpose.
Twenty meters down the avenue, some seventy-seven meters from
the center of the sarsen circle and about two meters off the center
line, is a single standing stone known as the heel stone. Over this
stone on the day of the summer solstice, the sun appears to
rise when viewed from the center of the Stonehenge complex
(Figure 18).

In front of the heel stone, as well as in the causeway entrance,
are the faint remains of many postholes. To archaeoastronomers, at
least, this suggests that sighting poles may have been once set here

either for astronomical observations or as preliminary tests of potential alignments before the final stones were put into place.

One of the popular misconceptions about Stonehenge is that the entire complex structure was built at the same time by the same people. In fact, Stonehenge was constructed in three distinct stages covering more than a thousand years by three peoples as different from each other in skill, intellect, and world outlook as modern Bostonians are from their Pilgrim predecessors.

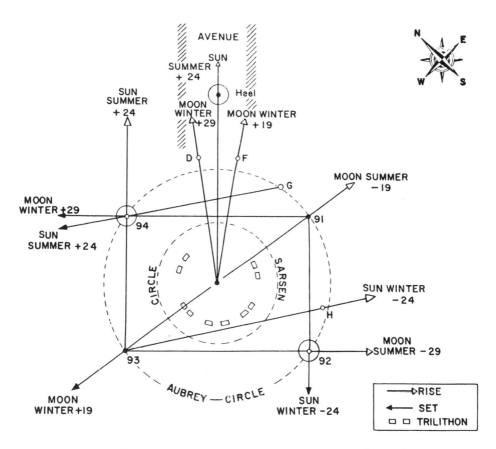

Figure 18. *Schematic diagram of the main alignments at Stonehenge showing the unique geometry of the site.* Illustration courtesy Smithsonian Astrophysical Observatory

The first phase of Stonehenge probably was begun around 2800 B.C. by a people known as the Windmill Hill culture because of their burial mounds found on nearby Windmill Hill. The Windmill Hill people were farmers and herders and often constructed crude cattle pens by digging ditches around the tops of small hills and then sticking wooden poles into the resultant dirt walls to form rough stockades. This same building technique was used at Stonehenge: A circular ditch was dug and the dirt thrown on a bank inside. An entrance gap about ten meters wide was left in the northeast segment. Outside the circle, but on a line with this gap, they erected a large stone. Just inside the ditch and earthen bank they dug fifty-six shallow holes; then, perhaps almost immediately, they filled them up with chalk and stone rubble. Later, and most significantly, at the four corners of a rectangle enclosed by the circle they erected stone posts or markers. Together with some scattered stones and a host of other postholes, including those in the causeway entrance, these features make up what is known as Stonehenge I, a structure contemporary with the Egyptian pyramids.

Perhaps four hundred to five hundred years later, somewhere around 2300 B.C., several changes were made in the layout of Stonehenge. The avenue leading away from the causeway was constructed, the ditch at the causeway was filled in, and the first double circle of bluestones was erected in the center. This last project was never completed, however, and was left with only about two-thirds of the stones erected. This was Stonehenge II, and the race thought to be responsible were the Beaker people, who were given this name from their custom of burying pottery as well as weapons and tools with their dead, usually in individual graves.

Sometime between 1900 and 1600 B.C., construction of Stonehenge III began, probably by the Wessex people, a Bronze Age society with a talent for both commerce and craftsmanship. The Wessex people erected the inner horseshoe of trilithons and the surrounding circle of thirty sarsen stones capped with lintels. The bluestone ring placed by the Beaker people was dismantled and the stones set in a new pattern: a circle inside the sarsen ring and a horseshoe inside the trilithons. In addition, the fifty-nine Y and Z holes were dug, for whatever reason. (Some archaeologists have argued that the "rectangle of station stones" was built coincidental with Stonehenge III and indeed served as the "stations" or benchmarks for laying out the final structure.)

Although the construction of the final phase required Herculean efforts by large and well-organized work gangs, the process was not as miraculous as many writers have suggested. The trilithon and sarsen stones were found as freestanding boulders on or near the surface of Marlborough Downs. They were dressed on that site by pounding with stone hammers. Using log rollers, the stones probably were dragged overland to Salisbury Plain by groups of no more than fifty to one hundred men, although some experts think as many as one thousand men would have been needed.

The bluestones, by contrast, were quarried from bedrock in Wales by a long and tedious process of alternately heating and cooling the rock until it eventually cracked. From the Prescelly Mountains, the bluestones could have been dragged (mainly downhill) to the head of the bay near the present-day Milford Haven and ferried around the south coast of England on rafts to the mouth of the Severn and perhaps even some distance up the Avon. Finally, they were dragged on rollers across the last few miles to Salisbury Plain.

At Stonehenge the stones were lined up around the circle with their bases pointing inward. A hole was dug under the base, while the other end was jacked up inch by inch using long poles as levers. Each time the stone rose a few inches, dirt and rocks were jammed under the rising end. Eventually the stone stood nearly upright and, with some strenuous nudging, it dropped base-first into the waiting foundation hole. With two stones standing side by side, a long sloping ramp of dirt and logs was built to the upper level and the third, or lintel, stone was dragged into place. The mortise-and-tenon joint for gripping and holding stones in place was borrowed from basic woodworking techniques of the period and actually represented a somewhat unnecessary redundancy in the building plan, since the weight of the lintel alone would have been enough to secure it.

Reenactments of this probable building and construction technique have been done successfully several times, using both scale and real-life models of stones. There was even one notable demonstration on British television employing scores of earnest Boy Scouts.

The building of the Stonehenge we see today took several hundred years, or longer than it took to build most medieval cathedrals. And unlike the builders of monuments who imparted to their work a sense of continuity provided by a single religious or ethnic background, Stonehenge's different work crews shared very little common cultural heritage. After ten or more generations had passed, the last contributors to the monument may no longer have even remembered why the project was begun.

Indeed, the final additions to Stonehenge—the sarsen circle and trilithon horseshoe so often identified in the popular and, to a certain extent, scientific mind with its use as an observatory—may have been made by peoples totally unaware of the significance of the earlier site. Harvard astronomer and historian Owen Gingerich, writing in *Technology Review,* suggests that "Stonehenge is not so much an ancient megalithic observatory as the monument to an earlier observatory. . . . Any astronomical sighting lines at Stonehenge must have been well established centuries before they were fossilized into such a heavy, immobile configuration, and the organization of the monumental stones is primarily dictated by the esthetic symmetry along their principal axis and not by a secondary series of lunar sightlines. . . . At best, Stonehenge was a ritual center commemorating bygone discoveries, not a site where new knowledge of the heavens was actively sought."

The true chronology of Stonehenge was not accurately known until the mid-1950s (and there are still some dating uncertainties); the earliest speculations about its creators—and its purposes—suffered greatly from serious misinterpretations of the evidence.

A possible astronomical use of Stonehenge was only one of many suggestions as to its function. By the end of the nineteenth century, every imaginable use had been proposed: Druid temple, monument to Merlin the magician, Roman forum, Viking parliament, Buddhist shrine, and even a gigantic gallows erected by the Saxons to hang defeated Britons. Modern interpretations have been no less bizarre. For example, Janet and Colin Bord feel the stones might have been giant storage batteries for both cosmic and terrestrial energies that could then be broadcast across the land.

It was the British antiquarian William Stukeley, actually in pursuit of the Druid connection, who made the first astronomical reference. In a work entitled *Stonehenge* and published in 1740, he hinted that the avenue pointed northeast, "where abouts the sun rises, when the days are longest." Following this lead, but also still believing in Druid origin, a Dr. John Smith wrote in 1771 that "from many and repeated visits, I am convinced it to be an astronomical temple." Smith drew a circle around the ditch, divided it into 360 equal parts, and then traced a line through the center to the heel stone to mark the sunrise at the summer solstice. Smith also believed Stonehenge to be an orrery, but it is unclear how he envisioned it marking the motions of the planets.

In 1880 Sir Flinders Petrie, one of Britain's greatest Egyptologists (who would later become involved in the controversy over astronomical interpretations of the pyramids), produced the first truly accurate survey of Stonehenge and made some calculations on the alignment of the structure with the solstice. Petrie was also the first British archaeologist to endorse the Stonehenge sunrise theory, writing that "only the first appearance [of the sun's disk] could coincide with the heel stone at any possible epoch." But it was Sir Norman Lockyer, an energetic, multitalented scientist, who made the first astronomical measurements at Stonehenge.

Lockyer came to Stonehenge in 1901. Previously he had measured Egyptian temples, claiming they were constructed to point at rising and setting stars and at the sun at the solstices and equinoxes. His book *The Dawn of Astronomy*, published five years earlier, gave the plans of dozens of Egyptian temples with their target stars identified. The book also developed a controversial theory linking the pantheon of Egyptian gods with the heavens. Working at Stonehenge with surveyor F. C. Penrose, Lockyer attempted to measure the midsummer solstice alignment and, from the mathematically known change in the obliquity of the ecliptic, put a tentative date on the construction of Stonehenge at 1680 B.C. (plus or minus two hundred years). This was reasonably close to the then current archaeological estimate of "Early Bronze Age." However, recent radiocarbon measurements show Lockyer's date was off by about 500 years. As mentioned earlier, modern archaeoastronomers frown on attempts to date a structure on the basis of suspected celestial alignments, preferring to leave the establishment of dates to archaeologists.

In measuring the summer solstice alignment, Lockyer also faced a problem that would plague all subsequent Stonehenge investigators: What features should one use to determine alignments? One can choose the midline of the avenue, or the axis through the center of the sarsen trilithon, or the heel stone as seen from one of several potential centers. Although the line of the middle of the avenue and the center of the stone circles seem to share the same azimuth to the horizon, the uneven shape of the ruined monument and the uncertainty in how closely the restored stones match their original locations actually makes any determination of the true axis or geometric center somewhat risky. Why should a few feet make a difference? When one uses two separated objects as sights for surveying a point on the horizon—just as one would use front and back sights on a rifle—and misaligns one sight by even a few inches, that error can distort by several hundred yards, or many degrees, the position of the observed phenomenon.

For his studies, Lockyer chose to use the center of the avenue rather than the heel stone, for that stone lies about 1.8 meters (6 feet) east of the midline. Today most archaeoastronomers feel he made the wrong choice. No matter, perhaps; his basic premise that Stonehenge was somehow linked with astronomy remains correct. (Recently Atkinson resurveyed the avenue and axis with modern instruments. He found that the azimuth was about one-half degree off from that cited by Lockyer. The exact cause of this discrepancy is not known, for nothing remains to show exactly where Lockyer set his stakes to mark the center line of the avenue. Nor are there any clues in the published records. Naturally, this would have been a serious error, and one that casts even further doubt on Lockyer's attempts to date the monument by archaeoastronomical means.)

While at Stonehenge, Lockyer also looked at the four station stones that form a large rectangle enclosed within the circle of ditch and bank. Lockyer was strongly influenced by the ideas about folklore and seasonal festivals expressed by his friend and colleague Sir James George Frazer in *The Golden Bough*. It was thus natural for Lockyer to seek markers of other ancient festival dates in addition to the midsummer celebration. Lockyer concluded that a line drawn from station stone 93 to stone 91 marked sunrise on or around February 7 and November 8. By reversing the line, looking now from stone 91 to 93, it pointed to sunset on May 6 and August 8. Lockyer felt that the stones marked the year's quarter days, the old calendar dates midway between solstices and equinoxes. Although imaginative, Lockyer's theory explained a purpose for only two stones out of nearly a hundred. Worse yet, his "quarter-day solution" may have blinded him to other, more interesting, relationships between the stones. As Hawkins has remarked, if Lockyer had not become so enthralled with the festival idea, he might have discovered the lunar alignment pattern linked to the rectangle created by the four station stones.

Lockyer went on to investigate other stone circles and monuments in Great Britain, all the while looking for clues in ancient myths and legends that might support the possible astronomical uses of various sites. Although one current trend in archaeoastronomy is to seek similar clues in the folklore of preliterate peoples, Lockyer's reliance on even the most dubious scrap of myth became both excessive and obsessive. Worse yet, he developed a diffusionist theory that suggested links between Britain and Egypt as early as 3600 B.C., as well as a theory that the Druids were really a Semitic people who had migrated to Britain.

Although much of Lockyer's work was uncontentious field research, his romantic excursions into the more fanciful aspects of prehistory provided fuel for his staunchest critics: the professional archaeologists. Lockyer was a distinguished scientist and the respected editor of the journal *Nature*, but he was not by training an archaeologist. Rather, his interest in astronomical alignments came about by chance, and with characteristic energy he pursued the study more vigorously than most of his archaeologist colleagues. This was not appreciated. Moreover, Lockyer's eclectic interests, sometimes sloppy scientific research, and overindulgent attitude to his own unreviewed articles in *Nature* did not win friends among archaeologists. Still, at the heart of their opposition was the simple resentment that Lockyer wasn't one of their gang. The same attitude a half-century later would haunt Hawkins.

Seven years after Lockyer's death, the archaeologist A. P. Trotter, writing in *Antiquity,* dealt the final blow. He dismissed Lockyer's work out of hand, calling into question the measurements, the axis alignment, and the idea of dating the monument by astronomical means. Trotter claimed the midsummer sunrise over the center of the avenue (or the heel stone) was pure coincidence. Somewhat testily he wrote: "We may prolong the axis to the northeast and find it hits Copenhagen, or ten and a half miles to the southwest to the village in which I live . . . and we may prolong controversies about it until we fill a library."

And there the matter ended. Lockyer was dead and so was the subject in the minds of most archaeologists. In the 1963 edition of the *Encyclopaedia Britannica,* Atkinson was able to write that the avenue was "aligned approximately on the point of the midsummer sunrise. . . . This fact has occasioned much fruitless conjecture." And, in his definitive book *Stonehenge,* Atkinson alluded to the supposed rising of the sun over the heel stone by saying: "It does nothing of the sort."

But denying it doesn't change the reality. Gerald Hawkins, born and educated in England, was aware of the folklore about the Stonehenge sunrise. He had walked around the site and had seen the published measurements of Lockyer. Hawkins found it difficult to reconcile the archaeological judgment with the observed facts. In his *Splendor in the Sky* (1961), he had written:

> There must be a great deal of magic that has been forgotten in the course of time. . . . Stonehenge probably was built to mark midsummer, for if the axis of the temple had been chosen at random the probability of selecting this point by accident would be less than one in five hundred. Now if the builders of Stonehenge had wished simply to mark the sunrise they needed no more than two stones. Yet hundreds of tons of volcanic rock were carved and placed in position. . . . Stone-

henge is therefore much more than a whim of a few people. It must have been the focal point for ancient Britons. . . . The stone blocks are mute, but perhaps some day, by a chance discovery, we will learn their secrets.

In 1960 while on a vacation trip to his home in England, Hawkins made a return visit to Stonehenge to do some preliminary surveying that might determine if Lockyer had been right or wrong. Because the positions of the moon and sun change very little over many centuries, Hawkins knew that a modern observation of the sunrise point could be easily extrapolated back in time to the epoch of 2000 B.C. And so he took the very simple and direct step of watching the midsummer sunrise at Stonehenge and taking photographs of the event for later calculations and measurements. He immediately found that the heel stone, when centered in the main sarsen archway, made a much better marker than the center line of the avenue. Even allowing for the elapsed years and a slight tilting of the stone, the rising sun would have appeared to "stand atop" the heel stone when the monument was first constructed.

At the time, Hawkins was chairman of the astronomy department at Boston University and a staff member of the Smithsonian Astrophysical Observatory across the river in Cambridge. For the Smithsonian, Hawkins supervised a joint program with Harvard University to measure by radio techniques the flux and velocity of tiny micrometeoroids in the upper atmosphere and to determine the hazard to the astronauts. Between his teaching duties and his meteoroid-hazard project, Hawkins had little time for hand calculations. Yet by the fall of 1960 he had determined the sun and moon alignments for the station stones as well as for holes G and H.

He was using as his standard survey map the fold-out plan in Atkinson's book, calibrated against his photographs of the sunrise over the heel stone. As he later told me, if he could have devoted full time to the project, he might have completed and cross-checked the calculations within about three months. In fact, his analysis fell far behind schedule and he finally turned to the observatory's computer system for help in speeding up the process.

Taking the best existing survey of the monument published by the British Ministry of Public Buildings and Works

(20-feet-to-the-inch scale), he determined the coordinates for 165 stone positions and fed them into a computer (initially, an IBM 705; later, a 7090) with instructions to (1) extend lines in both directions through 120 pairs of positions; (2) determine the true azimuths of these resulting lines; and (3) determine the declination—that is, the celestial latitude, at which these lines would meet the sky if extended to the horizon as seen from Stonehenge. At the same time, he used the computer to determine the significant horizon positions for the rising and setting of the moon, sun, planets, and stars between the years 2000 B.C. and 1000 B.C.

The computer produced a data set showing many duplications; in other words, a number of alignments showed the same declinations of the horizon: 29°, 24°, and 19° north; and their reverse, 29°, 24°, and 19° south. Hawkins checked the declinations against the positions of planets and found no correlations. Nor did he find any correlations with the stars, except a small number resulting from the chance alignment of two or more stones (and there were over 27,000 such possible alignments) with the vast number of bright stars in the sky.

With the stars and planets discounted as possible objects of alignment, Hawkins turned to the sun and moon. Obviously, he had suspected some correlations with the sun from the beginning of his study, but he was surprised by the number of new correlations shown in the computer printout. Not only did the heel stone line up with the 24° north declination of the sun at the summer solstice, but eleven other pairs of stones also aligned with this or other significant declinations (see Figure 19).

More surprising, however, the computer confirmed that the declinations of 19° and 29° north referred to the extreme declinations of the moonrise and moonset repeated in a nineteen-year cycle. (The reverse, or southern, declination aligned with winter positions of the sun and moon.) In fact, the computer showed 12 sun correlations with a mean accuracy of 1 degree, plus another 12 moon correlations with a mean accuracy of 1.5 degrees. The chance of these alignments occurring purely by chance was estimated to be less than one in a million, although mathematicians

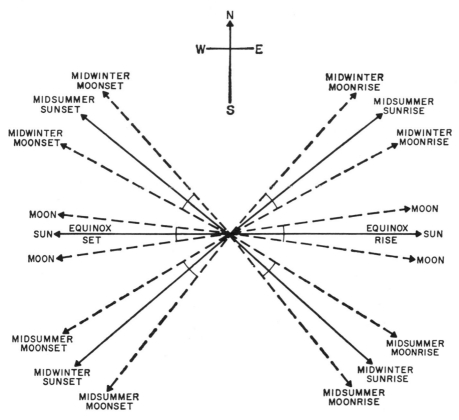

Figure 19. *The azimuths of the lunar and solar risings and settings as seen from Stonehenge on the solstices and equinoxes.* Illustration courtesy Gerald Hawkins and *Science*, vol. 147, pp. 127–30, January 8, 1965; copyright 1965, American Association for the Advancement of Science

are still arguing about how the exact probability could have been calculated. The preliminary announcement of his work was published in *Nature* (the journal founded by Lockyer) on October 26, 1963.

Unknown to Hawkins, another investigation of Stonehenge had been proceeding for several years and the results, in some ways similar to his own, had actually been published seven months earlier. A British amateur astronomer and retired gas-works engineer named C. A. (Peter) Newham had taken on the study of Stonehenge as a part-time retirement hobby. Beginning casually, he simply wanted to remeasure and resurvey the monument as a check on Lockyer's work.

But Newham's revised survey of Stonehenge showed a remarkable symmetry. A rectangle drawn using the station stones as the four corners would have its short sides parallel to the monument's main axis. Thus one side would point to the extreme northerly excursion of the sun, midsummer sunrise, while the other side, pointing in the reverse direction, would mark the midwinter sunset. At the same time, the long sides of the rectangle would point toward extreme excursions of the moonrise and moonset. Only at the latitude of Stonehenge will these directions form right angles (see Figure 20). According to Hawkins,

> From the standpoint of astronomical measurement, Stonehenge could not have been built further north than Oxford or further south than Bournemouth. Within this narrow belt of latitude the four station stones make a rectangle. Outside this zone the rectangle would be noticeably distorted. Perhaps these latitudes were deliberately chosen, and perhaps these people were aware that the angles of the quadrangle formed by the station stones would change as one moved north or south.

Newham found other tantalizing suggestions that certain features might be related to the significant moonrise positions. For example, he suspected that the station stone 92 and a slight depression in the turf just outside the Aubrey circle of holes, known as hole G on Atkinson's plan, might be aligned to the extreme northerly moonrise, one of four extreme positions in the nineteen-year cycle.

Newham sent his preliminary findings to Atkinson, the man whose book about the site had originally sparked his interest. Although it is unclear what his intentions were, Atkinson encouraged Newham and suggested that he submit an article to Glyn Daniel, editor of *Antiquity*, the British archaeological journal. Newham did so, and then waited more than two months for a reply. It finally came in the form of a rejection, with Daniel claiming that as a nonastronomer he could not properly evaluate the paper. (On the other hand, Daniel did not suggest sending it to another reviewer who might.)

Discouraged by his rejection in the scholarly press, Newham talked with a friend who was a science correspondent for the *Yorkshire Post*, a large provincial daily paper in Leeds. A story entitled

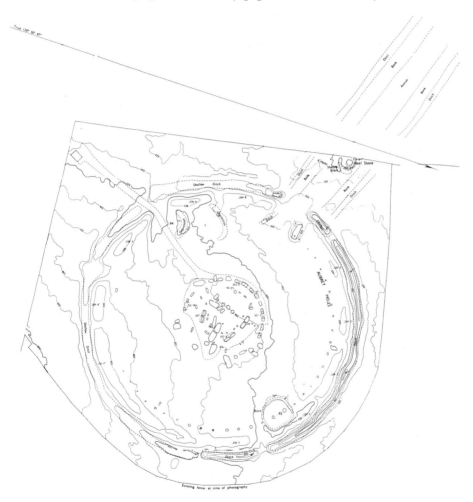

Figure 20. *Photogrammetric survey of Stonehenge prepared for the National Geographic Society and Smithsonian Astrophysical Observatory by Hunting Surveys, Ltd.* Illustration from *Beyond Stonehenge*, copyright 1973, Gerald Hawkins

"The Mystery of Hole G" appeared in the March 16, 1963, edition. The article showed how the long sides of a rectangle formed by the station stones might align with the moon's rise and set. The same point would form the heart of Hawkins's paper seven months later, although Newham chose directions in reverse to those calculated by Hawkins. Newham's observations would have been made from the mounds rather than over them, as Hawkins proposed. (A later photogrammetric survey showed that the sightlines actually would work both ways.)

Obviously, it was Hawkins who attracted world attention, while Newham's earlier publication went virtually unnoticed. But that attention was a mixed blessing. Certainly archaeologists did not accept the Hawkins theory with quite the same enthusiasm as did the press and public. Some, like Atkinson, were outspoken in their disbelief and opposition. Still, there was some cautious scholarly support. R. S. Newhall, the British archaeologist who had excavated at Stonehenge in the 1920s, said that he believed in its astronomical orientation.

Despite any disagreement among archaeologists, the new findings seemed to establish without doubt that there was some astronomical relationship to the geometry of Stonehenge. The monument definitely did point toward the northernmost, or midsummer, rising of the sun. No one could dispute that. On that dawn, the sun's disk can be seen to rise above the heel stone—or, more accurately, just slightly to its left. Four thousand years ago, the sun probably rose at least one degree north of this point, representing a slight change in the tilt of the ecliptic. The system still lines up on June 21 because the heel stone itself has tipped gently in just the right direction to compensate for the change in alignment. (Actually, the sightline from center to heel stone isn't really long enough to assure fine precision, so the sunrise appears to hover around the stone for almost a week before and after the solstice.)

The real question about Stonehenge—and the point of the continuing controversy over this ancient stone pile—is why it was built. Could Stonehenge have had any purpose beyond a simple "farmer's calendar" for marking the seasons?

Hawkins realized that astronomy might be only part of the answer. Lying as it does at the very center of a vast burying ground, Stonehenge must have also served as ritualistic temple or ceremonial meeting place. Yet because the monument would have been, as he described it, such "a dramatic backdrop for watching the interchange between sun . . . and moon," he searched for any other astronomical connections. For example, he wondered if there was some connection between the heavens and the various sets of numbers embodied in the circles of stones and holes at Stonehenge. Did the thirty sarsens or the twenty-nine Z holes represent simple lunar-period records? He also was fascinated by references in several local legends as well as by the first century B.C. writings of the Sicilian historian Diodorus Siculus. Diodorus's pseudoscientific history of the Hyperboreans related the moon and moon worship to Stonehenge. Diodorus also referred to the nineteen-year metonic cycle—the period in certain calendars after which the phases of the moon recur on the same days of the same months. The number nineteen appears again in the small horseshoe-shaped arrangement of bluestones, which has never been successfully explained.

When a sympathetic archaeologist asked Hawkins if something spectacular could occur at Stonehenge once every nineteen years, Hawkins thought of the slow swing of the winter moon across the width of the avenue every eighteen or nineteen years. He wondered, too, about a total lunar eclipse's occurring when the moon was visible over the heel stone or through the arch of the southwest trilithon. Although that structure marked the observational end-points of the lunar cycle, there apparently was no record of its period at the site. However, Hawkins felt that if the Stonehenge builders were interested in *where* the moon stood still, they should also be interested in *how long* it took to reach that point.

Hawkins knew from his already published work on the alignments that the rising point of the full moon at midwinter moved from its extreme northerly position to its extreme southerly position and back again, crossing the heel stone in the process. Similarly, the midsummer full moon swung across viewing lines in the trilithon archways. This usually took nineteen years; but, every third swing or so, the interval was a "short" eighteen years. This is because the regression of the nodes of the moon's orbit is not measured in exact solar-year units of 365.25 days. Thus, the moon's orbit returns to its starting point every 18.61 years. By using the computer to simulate what a Stonehenge priest might have seen visually, Hawkins found that, as close as he could tell, after nineteen winters the full moon would return again to the west side of the avenue. After nineteen more winters the view would be repeated; but then the alignment would show up after only eighteen years. In fact the most simple sequence would be three complete swings of the moon's extremes across the avenue; and 3 multiplied by 18.61 is 55.83, or very nearly 56.

Consulting a standard catalog of historical eclipses, and again using a computer, Hawkins could determine where and when eclipses of the sun and moon had occurred and were seen during the period 2000–1000 B.C. To his surprise, he found an eclipse of either the sun or the moon would always occur when the full moon closest to the winter solstice rose over the heel stone. (This occurred every nine or ten years at the midpoint of the moon swings, but the contact with the heel stone was still locked into the basic fifty-six-year cycle.) Hawkins knew, of course, that only about half of these eclipses could have been seen from Stonehenge. Still, based on anthropological data concerning the importance of eclipses for preliterate peoples, Hawkins felt the ability to be *prepared* for an impending eclipse could be as important as *predicting* it accurately. If a predicted eclipse did not occur, then the priest could take credit for preventing it. If it did occur despite various rituals, the priest could claim the sacrifices were not sufficient. In either case, the point was not to be *surprised* by the disappearance of sun or moon.

Hawkins further discovered that when the winter moon appeared over points D or F as seen from the center of the complex, then the harvest moon of that year would be eclipsed. The time between successive moon appearances over D was 18.61 years, the nodal period, and the harvest moon was eclipsed at intervals of nineteen or eighteen years, controlled again by the fifty-six-year cycle. Other moon alignments and eclipses also fit the same pattern. It was not a rigorous, long-persisting cycle like the famous saros cycle in which similar eclipses, either lunar or solar, recur at intervals of 6585 days. (This cycle was known to the Babylonians and will be discussed in more detail later.) Rather the Stonehenge cycle was a loose pattern that continued for about three centuries.

The recurrence of the fifty-six-year count throughout the pattern made Hawkins wonder if there could be some connection with the Aubrey circle of fifty-six holes. Could this circle have been a crude representation of time? The Aubrey circle had never been successfully explained, and it posed something of an archaeological mystery. The holes apparently had been dug a few years after the construction of the Stonehenge I ditch and then shortly thereafter were refilled with chalk and rubble. They had been variously described as postholes, ritual fire basins, and small graves for ceremonial interments. (The possible significance of the *number* of holes apparently had never been discussed, however.) Now Hawkins saw the circle as a huge counting device for predicting the movement of the moon and the "danger seasons" for possible eclipses. The results of his study were published in the article "A Neolithic Computer," in *Nature,* June 27, 1964. In it Hawkins explained how the fifty-six Aubrey holes could serve as a crude computer that allowed the astronomer-priests of Stonehenge to keep in step with the lunar cycles and to warn of possible eclipses.

By shifting a marker, perhaps a stone or carved wooden block, around the circle one hole per year, the positions of the moon could be tallied and the eclipse danger periods determined. The actual operation suggested by Hawkins was a bit more complicated in operation, however. Six markers would be used, three white stones and three black stones. Hawkins started his computer working at the time of a winter solstice lunar eclipse, when the full moon was at a node and had risen above the heel stone. At that time, a white marker was placed at hole 56, almost on the vertical axis of Stonehenge. The six markers were set out around the circle and spaced in a sequence of 9, 10, 9, 10, 9, 10 holes, alternating in color. The next year, with all stones moved counterclockwise by one hole, the first white marker falls in hole 55 and no winter solstice eclipse will occur.

After five years, the first white stone falls in Aubrey hole 51, and because the full moons near both the vernal and autumnal equinoxes are at the nodes, eclipses will occur soon thereafter. Four years later, a black stone falls in hole 56, and again an eclipse may occur. Any stone arrives at hole 56 at intervals of 9, 9, 10, 9, 10 years, and each time the arrival predicts the possibility of a lunar eclipse occurring on or near the summer solstice. A white stone falls in hole 51 with the sequence of 18, 19, 19 years, and predicts eclipses that may occur when the moon is at its greatest northern declination. A white stone arrives at hole 5 in a sequence of 19, 19, 18 years, and predicts a possible eclipse at the equinox associated with a 19-degree declination of the moon.

One of the major problems with Hawkins's computer, aside from its complicated record-keeping, was that it predicted accurately only a small fraction of lunar eclipses potentially visible from Stonehenge. At least one eclipse a year can be seen from any given position on earth, but the Stonehenge people would be prepared (if their computer worked) for only a selected few related to the most extreme lunar positions on the horizon. Of course these positions—and thus the eclipse potential—were already determined by the alignments of stones and posts, so the Aubrey-circle computer seems a little redundant. Hawkins later streamlined his computing system, reducing the number of moving stones to a single marker which was stepped around the circle three holes per year. Yet the complexity of his ancient computing system remained; and, while it never seemed to bother a technologically attuned public, it severely rattled the archaeologists.

The primary problem was that few British antiquarians could reconcile the apparently elegant and sophisticated computing techniques of the Stonehenge astronomers with their traditional view of these people as backward barbarians. Hawkins had suggested an awakening of intellectual ability nearly two thousand years earlier than existing texts (or lecture notes) indicated. The archaeological history of Britain would need reshuffling to account for this discovery.

Secondly, the popularity and general public acceptance of the computer-calendar explanation for Stonehenge seemed to rankle many British prehistorians. Hawkins's book *Stonehenge Decoded,* written with John B. White and incorporating both the original research and new supporting material, became a bestseller and inspired an extremely successful CBS Television special. This was not the usual public style of the British archaeological community. Certainly some of the more famous researchers of the past century had published books and explained their discoveries in the popular press; but, more often, the task of the popularization of science had been delegated to certain "approved" spokespeople. In the case of Stonehenge, the semiofficial voice of British archaeology would become Jacquetta Hawkes, who, together with Atkinson, launched a concerted defense of the traditional views of Stonehenge.

In an *Antiquity* article entitled "God in the Machine," Hawkes made the telling (if somewhat unfair) comment that every age had perceived the Stonehenge it desired. The romantic eighteenth century saw it as a Druid temple, while the Space Age public of the 1960s was overjoyed to find evidence of Stonehenge as a fore-runner of Palomar and Univac.

Atkinson did not comment directly on the original scientific ar-ticle in *Nature;* rather, he reserved his criticism for the Hawkins-White popular book. In an *Antiquity* review entitled "Moonshine on Stonehenge," he detailed all the archaeological errors made by Hawkins, even though the slips had already been corrected in the second edition—with acknowledgments to Atkinson. The title of the review was perhaps more fierce than the content, for Atkinson conceded that the Stonehenge designers must have "possessed a good deal of empirical knowledge of observational astronomy." In another review, written for *Nature,* he had called the book "tenden-tious, arrogant, slipshod, and unconvincing." And in the *Antiquity* review, he gave as an example of this arrogance the conclusion of Hawkins's chapter on eclipses: "I think there is little else in these areas that can be discovered at Stonehenge." Atkinson's selection of the quote was a bit ungenerous, since the entire citation read

I think there is little else in these areas that can be discov-ered at Stonehenge—although I must confess, as I make that flat statement, that I am filled with trepidation, and cannot forget how often the old monument has produced new astonishments.

The machine has established an extraordinary sun-moon correlation throughout the structure. Astronomy has done its best. It now rests with the prehistorians, the archaeologists, anthropologists, mythologists and other authorities to make use of these new findings to advance our understanding of the "gaunt ruin," which should no longer stand *quite* so lonely in history as it does on the great plain.

The controversy that followed in the scientific and popular press has been excellently reported by Peter Lancaster Brown in his book *Megaliths and Masterminds,* as well as by Hawkins in his own *Beyond Stonehenge,* which he views as his rebuttal to the arguments.

Hawkins was not alone, however, in defending the astronomical interpretations of Stonehenge. C. A. Newham continued his understated and careful studies of the site, publishing a handbook privately. Before his death, he completed a study of the lunar alignments of the postholes in the avenue; the computations were done for him by Hawkins using the Smithsonian computer. That paper was accepted by *Nature.*

Hawkins also received support from an unexpected ally: Sir Fred Hoyle, one of Britain's most distinguished astronomers, who published his own paper in *Nature* on July 30, 1966. No matter what any archaeologists felt, said Hoyle, Stonehenge was an astronomical observatory and the Aubrey holes were part of an eclipse-predicting system. They just didn't operate quite the way Hawkins suggested. Hoyle then proceeded to describe his own method for operating the eclipse predictor; and, while even more esoteric than Hawkins's technique, it did predict all eclipses. (Hawkins, of course, had limited his hypothesis to the "danger periods," not eclipses per se.)

Hoyle interpreted the Aubrey circle to represent the ecliptic, and by moving markers about the circle to represent the moon, sun, and nodes, he showed that an observer could predict nearly every lunar eclipse on earth with some accuracy, even though only half would be actually visible from Stonehenge.

Hoyle's explanation, like much of his cosmological theory, was so dense, so mathematical, and so much more complicated than Hawkins's that archaeologists were again unable to reconcile the apparent savagery of the early Britons with the sophistication of the system. However, Hoyle himself made a vital statement that should be remembered in all discussions about archaeoastronomy:

> The eclipse interpretation requires an apparently subtle piece of astronomical information: that in the month of the 18.61-year-cycle when the Moon swings most to the north, the Node lies behind the midsummer position of the Sun by 90 degrees. But this information is subtle only if one is required to understand it. No subtlety is required to arrive at this fact by direct observation. . . .

Debate on the ability of prehistoric man to compute eclipses will continue, no doubt, through the lifetime of the current debators and beyond. To the nonmathematical aficionado of Stonehenge, the debate over the prediction system is sometimes mind-numbing. As Hawkins admits, "It's unfortunate we need the mathematics, but the math does let us go back in time. Although the computer has been my time-machine, I would much rather go back in person." But the eclipse question is no mere side issue. "If the eclipse interpretation of Stonehenge be admitted," Hoyle wrote in *On Stonehenge*, "the intellectual achievement of the builders must stand out in prehistory like a veritable Mount Everest."

Still, for many observers—even the most sympathetic—this accomplishment is almost too incredible to accept. On the other hand, as science historian Owen Gingerich, an objective critic of the Hawkins-Hoyle computer theory, wrote in *Astronomy of the Ancients:*

> For a people so concerned with capturing the northernmost position of the sun and moon, the conception of the moon's

nodes may not be terribly far behind. In other words, it may well have been possible for the Stonehengers to have correlated eclipses with the celestial geometry of the solar and lunar paths rather than with cycles of eclipses derived from a communal memory of events long past. . . . This seems to be a fabulous jump for neolithic man to have made, but there is nothing to have prevented a Stone Age genius from finding the correlations simply with sticks and stones.

Stonehenge, of course, is not the only great Neolithic monument in Northwest Europe—far from it! Although the best known, Stonehenge is only one of literally thousands of rings, standing stones, grave sites, and earthen mounds dotting the British Isles and the coast of Brittany. Indeed, the area immediately surrounding Salisbury Plain itself is so covered with similar structures that much credence was given to early theories that this was some sort of prehistoric fairground where tribes gathered for ritualistic games culminating in ceremonies at the main temple of Stonehenge. Although only scant connections can be found between specific sites and Stonehenge, an increasing body of evidence suggests that many other structures had astronomical alignments, thus supporting such interpretations of Stonehenge.

Lockyer ranged far beyond Stonehenge in search of astronomical alignments; and, later, Vice Admiral Boyle Somerville followed up his work at some twenty-seven sites, including a major complex at Callanish in the Outer Hebrides off the coast of Scotland.

The most diligent and devoted successor to Lockyer, however, must be Alexander Thom, a retired Oxford professor of engineering science, who for nearly thirty years has been making accurate surveys of major sites in Great Britain.

Thom's method is to visit megalithic sites and, using the surveyor's theodolite and tape, measure the positions of all stones as accurately as possible. His work, long and plodding and extremely solitary, attracted little public attention and was generally ignored by archaeologists until after the Stonehenge affair hit the press. Most of the sites identified by Thom as "prehistoric precision observatories" incorporated features that used backsights marked by standing stones and some sort of distant foresights marked on the far horizon by a mountain notch or peak.

Thom's research was of great relevance to the Stonehenge findings, for it suggested that an astronomical tradition was widespread throughout prehistoric Britain. Thom had already voted for archaeoastronomy. Writing in *Antiquity,* he said he was "prepared to accept that Stonehenge was a solar and lunar observatory." Moreover, Thom's survey of Stonehenge in 1973 showed that the lunar and solar sightlines might be extended from the monument stones to distant points on the horizon for greater accuracy, thus supporting the astronomical interpretation.

His work resulted in the publication of several papers, including one in the *Journal of the British Astronomical Association* entitled "The Solar Observatories of Megalithic Man." Thom attempted to show that the megalithic builders of Britain had a much finer grasp of geometry than any prehistorians had previously imagined, utilizing the properties of the ellipse in the design of compound circles. Thom also became convinced that ancient man had developed a standard unit of length—which he called the megalithic yard—that influenced the shape and dimension of the stone circles, many of which are really distorted into egg shapes and ellipses.

The "megalithic yard" was at the heart of Thom's theories. He placed this measure at 2.72 feet, with the so-called "double-yard" at 5.44 feet. Only with the existence of such a standard measure guiding the Neolithic architects do many of Thom's sites make geometric sense. His analysis of the various stone monuments, including the giant Avebury circle near Stonehenge and that enigmatic collection of standing stones at Carnac in Brittany, are extremely complex, but the crux is Thom's assumption that the megalithic builders were unsatisfied with the uncertainty in the value for pi. He feels they intentionally used another, more approximate, number that resulted in distorted circles but fulfilled their sense of perfection in mathematics.

The second part of Thom's work was to show that many stone circles also had solar and lunar alignments. Using the standing stones of the circles as backsights, he traced alignments to the rising and setting moon and sun over distant horizon features, sometimes as far as eighteen to twenty miles away. The apparent accuracy of Thom's measurements impressed some archaeologists who had previously resisted astronomical interpretations. Even the reluctant Atkinson eased his position on the astronomical uses of Neolithic sites and grudgingly conceded that such alignments were possible at Stonehenge, too.

Unfortunately, in recent years the work of Thom has undergone some reappraisal by his fellow archaeoastronomers—and has been found lacking. Evan Hadington's field research at several major "observatories" cited by Thom, including Ballochroy, Kintraw, and Carnac, reveals the alignments to be really quite crude. Although they may still be astronomical, the alignments conflict sharply with Thom's hypothetical models of high-precision geometry. Other critics have noted with dismay the "saw-toothed horizons" used by Thom. In other words, the horizon line is so dotted with possible foresights that almost any number of astronomical alignments are possible; one merely need choose the horizon feature that best fits.

Aubrey Burl, another well-known investigator of the stone circles, suggests the structures may have been more symbolic than scientific. In almost all of the hundreds of "recumbent stone circles," those in which at least one stone of the circle lies flat on the ground, he found that the recumbent slab points south. While this may indicate an orientation toward a lunar or solar position, he claims the orientation "cannot possibly be used for any astronomical prediction whatsoever." Rather, he interprets the orientation toward astronomical objects as a general feature of ancient burial practices. For example, the east–west orientation of many graves seems to have a relationship with the natural life-and-death cycle symbolized by the rising and setting of the sun.

In the case of the recumbent stone circles, he thinks the gap in the ring was deliberately planned so the moon's rays might enter the circle over this flat stone. He even suggests that when the summer moon appeared low on the southern horizon, its light was reflected off pieces of quartz scattered about the inner circle. According to Burl, since these structures were tribal or clan ritual centers, stones may have been broken or removed when one clan attacked the circles of another.

One of the best examples of a stone ring with possible astronomical implications is Callanish on the Isle of Lewis in the Outer Hebrides. Located on the western side of the island not far from

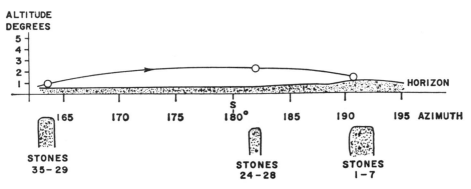

Figure 21. *The horizon-skimming of the full moon as seen from Callanish at midsummer in about 1500* B.C. Illustration courtesy Gerald Hawkins and *Science*, vol. 147, pp. 127–30, January 8, 1965; copyright 1965, AAAS

Stornoway, and overlooking East Loch Roag, Callanish has the rough shape of a Celtic cross. The long axis, pointing north–south, is 405 feet long; the east–west axis is 140 feet long. The central circle has a diameter of 42 feet and is made up of 13 stones, ranging in height from 8 to 12 feet. Within this circle are the ruins of a chambered cairn, and cremated human remains were found here some 120 years ago. Five stones on a line running south point toward a rocky outcropping known as Croc an Tursa. Four stones form each arm of the cross, and a double row of nineteen stones forms an avenue that points just east of true north. The largest stone in the complex stands by itself in the center as part of the cairn; and two other standing stones are outside the center ring, one to the southeast, the other to the southwest (see Figure 23).

Like Stonehenge, various theories have been suggested over the centuries to explain Callanish: a barrier against ghosts, a field of petrified people, Viking gravestones, Druid temple, Greek shrine to Apollo, open air parliament, fertility temple, Christian monument, an exercise in advanced geometry, and, most significantly, as astronomical observatory using movements of the sun to establish planting and harvest times and to predict eclipses.

Encouraged by the last theory, Lockyer studied the plans of Callanish (he never traveled there personally) and developed a sightline down the avenue that matched the visibility of the star

Capella in 1720 B.C. He also found that an alignment from the cairn could have coincided with observations of the Pleiades in 1330 B.C.

Somerville continued Lockyer's work and actually visited the site in 1912 to draw up a detailed plan. On the basis of his survey, he found what he thought were several significant alignments: the avenue with the rise of Capella in 1800 B.C.; the east-row sightline with observations of the Pleiades in 1750 B.C.; the use of the west row with sunset on the equinoxes; and the offset stone on the southwest and the first stone on the northeast edge of the avenue with the extreme midwinter rise of the full moon.

If Stonehenge was built at a significant latitude for solar observations, then Callanish seems equally near a significant latitude for watching the moon. Indeed, every 18.61 years the moon rises so close to due south that for a few days its path is less than 2 degrees above the horizon. The full moon literally skims across the southern skyline (see Figure 21). Knowing this, it would seem Diodorus of Sicily, writing about the Hyperboreans in his *History of the Ancient World*, might have been describing Callanish when he wrote:

> There is also on the island a notable temple which is spherical in shape. The moon when viewed from this island appears to be but a little distance from the earth. The god visits the island every 19 years and dances through the night from the vernal equinox until the rising of the Pleiades.

Hawkins, looking at Callanish for further support of his Stonehenge theories, applied the computer to an analysis of azimuths through combinations of stones and found twelve possible alignments with moon and sun rises and sets (see Figure 22). Three alignments with the extreme southern positions of the midsummer moon hinted at some connection to Diodorus's history. Callanish lacked any features similar to the Aubrey holes of Stonehenge and Hawkins suggested the thirteen stones could have been used as a counting system to establish a lunar calendar. However, he considered this aspect as conjectural because of the uncertainty of the missing stones.

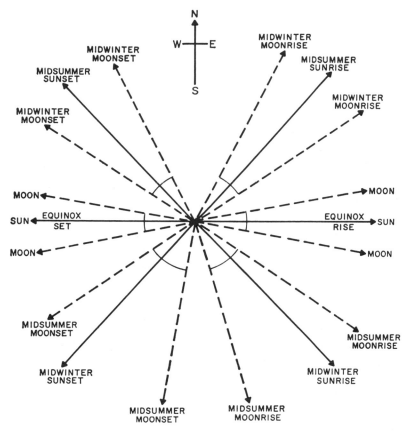

Figure 22. *Azimuths of solar and lunar risings and settings as seen from Callanish on the solstices and equinoxes.* Illustration courtesy Gerald Hawkins and *Science*, vol. 147, pp. 127–30, January 8, 1965; copyright 1965, AAAS

Thom devised a similar theory. Using Somerville's detailed maps, Thom concluded that the central stone ring was really a flattened circle, so that a point just south of the central cairn marked the meeting of lines drawn from the east and west rows and along the central path of the avenue. From this point, an alignment with a horizon feature along the axis of an ellipse would mark the midsummer sunrise; another alignment along the avenue could mark the rise of Capella (supporting Somerville) and the moonrise to the south (supporting Hawkins). He further suggested that the slightly

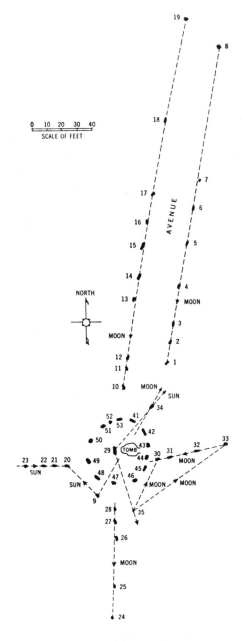

Figure 23. *A site plan of Callanish with alignments to sun and moon points.*
Illustration courtesy Gerald Hawkins and *Science*, vol. 147, pp. 127–30,
January 8, 1965; copyright 1965, AAAS

skewed construction of the stone rows was done deliberately to mark these astronomical alignments.

Unfortunately, a 1974 survey by the University of Glasgow's Geography Department (and a photogrammetric aerial survey made by Hawkins in 1965) revealed that Somerville's original plan was inaccurate in several respects. Moreover, although Thom had noted the extreme southerly position of moonset over the Clisham Mountains to the south, he had worked from contour maps rather than field studies. He had not realized that the rocky outcropping of Croc an Tursa hides the view to the south from Callanish.

The point of mentioning this error in calculation at Callanish is not to demean the contributions of Thom, but merely to show how even the most meticulous and precise researcher can make mistakes in the search for astronomical alignments. Such errors continue to provide fodder for the critics of archaeoastronomy. As R. J. C. Atkinson wrote in *Nature* in 1977,

> From our present knowledge of celestial mechanics it is all too easy to impute to our prehistoric ancestors an understanding of nature which may well be ours alone. Concentrating exclusively on the astronomical interpretations of aspects of a monument which was surely more than an observatory (if it was ever that), leads to invalid inferences and to the posing of non-problems . . . One need look no further than many Christian churches oriented on the sunrise either at the equinoxes or the patronal Saint's Day to see that a religious purpose need not be compatible with some elements of astronomy; nor does the incorporation of sundials and clocks in churches imply that they were built primarily as timepieces.

Atkinson may still be missing the point—and the greatest benefit—of archaeoastronomy. Hawkins, Hoyle, and Thom, as well as their many colleagues and their growing number of converts, are not saying that all prehistoric men were budding Einsteins or even full-fledged astronomers whose "professional" lives were attuned to the stars. Rather they are suggesting several possibilities, none incompatible with traditional archaeology: (1) Prehistoric man had the intellectual capacity to make basic observations of natural phenomena; (2) these same people could develop tools, admittedly crude and cumbersome, to aid in the observing and recording; (3) astronomy played an important, indeed central, part in their lives but it was fully integrated, not a specialized activity such as we associate with today's academic pursuit of science; preliterate astronomy was practical, immediate, and direct; and perhaps most important, (4) the experience of watching, recording, measuring, and predicting led to the development of cognitive abilities that may have laid the foundation for the development of other technologies.

This last point is valid even if the basic knowledge of megalithic monument building may have gone into a long period of dormancy. For example, the astronomical orientation of Stonehenge I is almost beyond dispute, but its relationship with other stages of Stonehenge remains unclear. It is possible that after 1700 B.C. the climate of Great Britain deteriorated, changing from a land of clear skies and cloudless horizons to an island of fog and rain, thus making observations more difficult. As the observations declined, the old astronomical knowledge may have been lost. As Owen Gingerich argues, the Stonehenge we know and recognize as the familiar circle of great rocks might merely be the monument to an earlier observatory. Other sites, too, including those oddly shaped rings investigated by Thom, may be the mere echoes of an earlier and transient brilliance. Observational astronomy may have been replaced by ritual, with the old alignments and shapes roughly maintained by priests with only vague memories of their original purpose.

But even this development is important for archaeologists, or for anyone attempting to understand humankind. Throughout history ideas and concepts have sprung up, developed to a peak, and then slowly declined as they lost relevance to the society that created them. "With this in mind," writes British engineer and archaeologist John Wood in *Sun, Moon and Standing Stones,*

it should be possible to trace the evolution of geometrical and astronomical ideas and relate them specifically to other aspects of life. There must, for example, have been some evolutionary process which led to the different shapes of the stone rings; and, ultimately, it is to be hoped, when the chronology of stone circles is better known, each type of ring will be related to its own time span and geography, and connected with other cultural features. Eventually, the shape of stone rings, which now have no clear pattern in time or space, could become one of the clues by which we recognize the cultural characteristics from one community to another. Similarly, it ought to be possible to reconstruct the development of astronomical techniques in detail. . . . We now have a prospect which is unique in archaeology, and particularly attuned to our own time. We are discovering the first traces of the development of science in one of the most remarkable periods in the whole of man's history.

Star Rise
Over the Nile

"The main difficulty which Egyptologists face is the recreation of a state of mind of human society 5,000 years ago. [However], although man's spiritual world-picture and his moral laws have changed out of all recognition, the laws of physics remain unaltered. The knowledge that these same laws were operative and had to be obeyed 5,000 years ago in exactly the same way as today provides a reliable link between the pyramid builders and ourselves."

—Kurt Mendelssohn

The guide hiked up the skirts of his flowing robe and, with the condescending efficacy of guides everywhere, wordlessly scrambled up the rocky north face of the pyramid of Khefren at Giza. My wife and I, panting and sweating under the late-afternoon Egyptian sun, struggled after him as he popped through a gap in the stones. Once inside this passage entrance, we felt the temperature drop dramatically. The streams of perspiration running down my face and back immediately turned to icy ribbons—partly from the cold and partly from the realization that some five million cubic yards of limestone were now poised above our heads (see Figure 24).

Unlike the Great Pyramid of Cheops, the entrance to the subterranean chamber of Khefren's pyramid is fairly direct. The first hundred feet drop sharply at a steep angle to join a second en-

Figure 24. *The pyramid of Khefren at Giza. Note the limestone cap still in place.*

trance passage rising from a point near the base of the pyramid, but the remaining two hundred feet of the passage are nearly horizontal. Although high and broad enough for persons of normal height to walk comfortably upright, one still has an uncontrollable urge to stoop, indeed crawl. The walls and floor of the passage are remarkably dry; but, again, one draws back and avoids touching the rock. No need to tempt fate! A weak yellow glow from a widely spaced string of naked electric bulbs illuminated the passage. The guide, keeping up a pace developed over thousands of similar trips in which he had learned to increase his volume of tourists through the brevity of his visits, faded from view down the passage. We ran, crouched over, trying to keep him in sight.

We were alone in the passage. At this time in early 1973, between successive Egyptian-Israeli conflicts, there were few foreign visitors at Giza. Our aloneness was heightened by the faint sound that echoed down the empty passage. An irregular clicking noise seemed to be coming from far away, somewhere ahead of our hurrying guide. It was exactly the kind of noise one would expect in some terrible B movie about the mummy's curse. A giant scarab beetle, perhaps, chirping a warning to graverobbers? The rattle of ancient bones? Or the rhythmic opening and closing of a coffin lid?

Suddenly the guide stopped. We entered the large chamber dug into bedrock nearly dead center beneath the pyramid. One end of this chamber looked as barren and empty as when the first grave looters reached it three thousand years earlier. The other end was boarded over by a crude plywood wall, and the clicking sound came from behind this partition. The guide knocked on the partition and a door opened slowly. No linen-wrapped mummy greeted us, but rather a short, stocky man with graying hair, wearing the international uniform of the electronics engineer: dark gabardine pants, sensible shoes, and a short-sleeved white shirt, its breast pocket crammed with pens and pencils in a protective plastic sheath. Behind him, brightly lit with fluorescent bulbs, was a self-contained physics laboratory. A large spark chamber clicked loudly as cosmic rays passed through its alternating leaves of metal and gas to produce electrical discharges duly recorded on a nearby oscilloscope and digital punchtape machine (see Figure 25). Here, 470 feet below the pinnacle of Khefren's pyramid and some 9,000 miles away from home, Nick Chakakis of Berkeley, California, was using the latest techniques of modern astrophysics to decipher one of the most puzzling mysteries of archaeology: Where was the tomb of Khefren?

Built in approximately 2600 B.C. by the same Fourth Dynasty pharaoh whose features are seen on the face of the Sphinx, the second largest pyramid at Giza has always posed a problem for Egyptologists. Overshadowed by the sheer size of the Great Pyramid built by Cheops, Khefren's pyramid has fascinated researchers more by what it lacks than what it has. Most of the other pyramids, and especially Cheops's, contain a maze of hidden halls and rooms, ingeniously constructed to foil graverobbers. Cheops's pyramid, for example, has an elaborate underground chamber, plus upper chambers built into the masonry and guarded by an array of gran-

Figure 25. *University of California at Berkeley technician Nick Chakakis tending a spark chamber located deep within the pyramid of Khefren during a search for the "missing tomb."*

ite plugs, blind alleys, concealed passages, inexplicable air shafts, and stone portcullis gates—some set to collapse if disturbed. All of these complicated structures, complete with dummy coffins and sarcophagi set out in the more easily accessible chambers, were apparently designed to confuse potential looters, making them believe the tombs had already been plundered. (The ploy did not work, of course, and the pyramids were sacked anyhow.)

In Khefren's pyramid, however, there is only one chamber, connected almost directly with the outside. The secret passages and false turns are not here, but neither is there a master burial chamber (see Figure 26). This has never made sense to Egyptologists. Why would Khefren, who had the experience of his pyramid-building father, Cheops, and grandfather, Sneferu, to guide him, take so little care to hide his remains? For centuries researchers have been convinced that somewhere within the pyramid there must be a master chamber—a chamber filled with treasures to rival anything found in King Tut's tomb. However, all searches have been in vain.

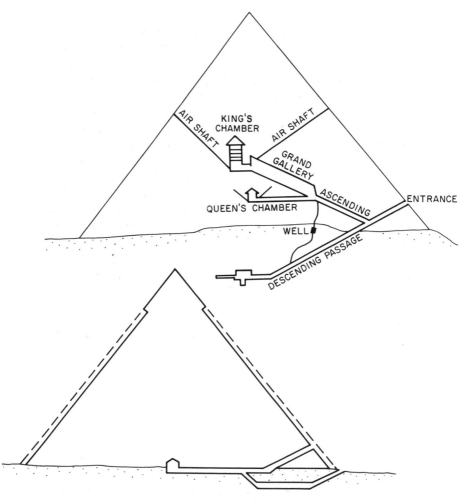

Figure 26. *Cross-sections of Cheops's pyramid* (top) *and Khefren's pyramid* (bottom). *The apparent simplicity of the passageways and the single chamber in Khefren's pyramid has led to much speculation about a hidden "tomb" somewhere within the structure.* Illustration by Joseph Singarella, Smithsonian Astrophysical Observatory

In the late 1960s the Nobel Prize–winning physicist Luis Alvarez of the University of California at Berkeley devised a plan to "X ray" the millions of tons of limestone and locate the lost chamber. Alvarez, whose specialty is the study of cosmic radiation, proposed that by measuring variations in the flux of energy passing through the pyramid he could determine if any hollow spaces existed within the superstructure.

Cosmic rays are extremely high-energy particles generated in the cores of distant stars and then sent traveling throughout the universe. When these rays collide with the earth's atmosphere, they create a shower of subatomic particles that in turn rain down on the surface of the earth with known energies and at known rates. One particle, the mu meson, or muon, is so energetic that it penetrates almost all material to depths determined by the muon's energy and the material's density. (Thousands of muons pass through our bodies every day.) If Alvarez placed a "muon counter," or spark chamber, beneath the pyramid and measured how many muons passed through the material above, he reasoned that he should be able to tell if the pyramid was solid.

"We would expect a given flux of muons to strike a given target area," said Alvarez. "The rocks of the pyramid would subtract some; but a hole in the rock would, in effect, add some back. A fifteen-foot hole in a three-hundred-foot pyramid, for example, would give a ten percent increase in intensity."

With the cooperation of the Egyptian government and a grant from the Smithsonian Institution, a muon counter was placed in the known chamber and laid flat so it scanned an area directly through the center of the pyramid. The first effort X rayed about 20 percent of the pyramid in a cone-shaped section that had its point in the chamber and encircled the peak. Indeed, Alvarez and his Egyptian colleagues were able to detect the extra thickness of the limestone cap that still remained on the apex of this pyramid and to identify precisely the distance by which the burial chamber is located off center. They did not find a missing tomb before political problems in the Middle East halted further study.

By 1973, however, when my wife and I visited Nick Chakakis in his subterranean workshop, the project, now supported by the National Science Foundation, had been reinstated and redesigned so the spark chamber could be tipped at an angle parallel to the pyramid's slope. Chakakis spent several months on the site pointing the detector at the corners and sides of the pyramid and scanning the remaining 80 percent of the pyramid. The results were conclusive. Alas, they were also negative; the scan did not reveal any lost chambers. In a sense, Alvarez's experiment only increases the mystery. Why did Khefren bother constructing a pyramid-tomb, if he never intended to use it? Not even modern physics can answer that.

It is fitting, perhaps, that modern astronomical techniques should have been applied to solving the secrets of the pyramids. For centuries a dedicated and sometimes fanatical band of investigators, or "pyramidologists," have ascribed all sorts of purposes and functions to the pyramids, and astronomy has always remained one of the most popular (see Figure 27). But before tackling the particular—and often diversionary—problems raised by the pyramidologists, we should examine the real role astronomy played in ancient Egypt.

The ancient Egyptians, like the Babylonians who will be discussed later, are popularly considered to have had highly advanced astronomical knowledge. The Egyptians' reputation may be somewhat unjustified. In fact, the Egyptians made very little noteworthy progress in the astronomical sciences. As David Allen wrote in a *Sky and Telescope* article,

> The ancient Egyptians kept copious records of their daily affairs, but left little indication of significant astronomical interest. That they were impressed by the sky, the moon, planets, and stars, is obvious and this would be expected for a people living in a desert environment. Yet there is little evidence that such matters played a significant role in their thinking.

Otto Neugebauer, author of the classic work *The Exact Sciences in Antiquity*, was describing ancient civilizations in general when he said that mathematics and astronomy had practically no effect on the realities of life, but he might also have been describing the Egyptians in particular. "The mathematical requirements for even

Figure 27. *A fanciful nineteenth-century view of a comet over the Great Pyramid.* Illustration courtesy Smithsonian Astrophysical Observatory

the most developed economic structures of antiquity can be satisfied with elementary household arithmetic," he wrote, "and the development of ancient astronomy was relegated to the status of an auxiliary tool when the theoretical aspects of astronomical lore were eventually dominated by their astrological interpretations."

The modern archaeoastronomer, however, is not concerned solely with theoretical applications when he searches for evidence of astronomical observations among ancient peoples. He is most concerned with the emergence of astronomical awareness, not whether it eventually blossomed into a science or a ritual. Although the priests of ancient Egypt controlled astronomy and jealously guarded their observational results for use in astrological predictions, some very practical applications of astronomy were also made.

From the time that the first humans settled in the Nile Valley, the periodic event of prime importance to their lives—their very survival—was the annual flooding of the river. This cyclic event, crucial to the establishment of Egyptian civilization, also led naturally to the concept of time. The need to prepare for the flood each year required the development of a calendar.

The first calendars were based on lunar observations, and a careful noting of the last waning crescent visible in the predawn sky would give a fairly accurate period of 29.5 days. Unfortunately, a synodic year, one based on multiples of monthly lunar periods, does not quite match the tropical year, the time from one summer solstice to another. And so as early as 3000 B.C., the Egyptians had adopted a solar calendar of 365 days, divided into 12 months of 30 days each, with an extra 5 days tacked on to the end.

As the starting point of this year, the Egyptians chose the heliacal rising of Sirius, the brightest star in the northern hemisphere. The heliacal rising, as you will recall, is the first day on which a star can be seen to rise briefly ahead of the sun in the morning twilight. By happy coincidence, at the epoch of ancient Egypt Sirius first appeared in the morning sky around the summer solstice and at about the same time as the start of the Nile flood. No doubt the Egyptians saw this coincidence as supernatural, for they identified Sirius with the goddess Isis. The length of the Egyptian solar year was thereby set as the interval between the successive risings of that star. Of course, the solar year was still too short, and the calendar lost one day every 4 years and a month every 120 years. Worse yet, any divisions of this solar year into equal parts soon had the "civil" or "religious" calendar out of phase with the real seasons.

The Egyptians realized the need for some sort of fixed calendar unrelated to the solar year which could set dates for certain festivals. Knowing that the wandering solar year lost a day every four years, the priests simply correlated the wandering year with the fixed year set by Sirius and established dates for all past and future times. By extrapolating into the future, they found that the beginning of the wandering year would again match the beginning of the fixed year after 1,460 years of heliacal risings. This period became known as the Great Year of the Sothic cycle (from *Sothis*, the Egyptian name for Sirius).

The religious and civil calendar used by the Egyptians from 2782 B.C. until Roman times was based on a wandering year of 365 days, and holidays and festivals simply drifted slowly through all the seasons during the 1,460-year cycle. By contrast with modern calendars, which have months of differing lengths and the intercalation of days every 4 years and every 400 years, the Egyptian calendar was the model of simplicity and uniformity.

In fact, Neugebauer, in his *Exact Sciences in Antiquity,* called the Egyptian system

> . . . the only intelligent calendar which ever existed in human history. Though this calendar originated on purely practical grounds, with no relation to astronomical problems, its value for astronomical calculations was fully recognized by the Hellenistic astronomers. Indeed, a fixed time scale without intercalation was exactly what was needed for astronomical calculations. The strictly lunar calendar of the Babylonians with its dependence on all the complicated variations of the lunar motion, as well as the chaotic Greek calendar, depending not only on the moon but also local politics for its intercalation, were obviously inferior to the invariable Egyptian calendar. It is a serious problem to determine the number of days between two given Babylonian or Greek New Year's Days, say 50 years apart. In Egypt, this interval is simply 50 times 365. No wonder the Egyptian calendar became the standard astronomical system of reference which was kept alive through the Middle Ages and was still used by Copernicus in lunar and planetary tables.

The second major contribution of the Egyptians to astronomy—and to our modern civilization—was the division of the day into 24 hours, 12 of daylight and 12 of nighttime. Evening hours were measured by observations of stars. (The night sky was divided into 36 decans, each rising for a period of 10 days, thus providing a count for the passage of days as well as hours.) Daylight hours were counted by sundials. (The first known Egyptian sundial dates from 1500 B.C. It was simply a flat stick with a T-bar at one end. The times of day were deduced by the position of the shadow cast by the crossbar onto marks inscribed on the long stick. The crossbar would be turned to face east in the morning, west in the afternoon.) For the Egyptians, the twenty-four hours were not equal; they varied in length depending on the season, with daylight hours longer in summer and shorter in winter. The Greek astronomers later replaced these "seasonal hours" with "equinoctial hours" of equal length. Because at this time all astronomical computations were based on the Babylonian-inspired sexagesimal system, the hours were divided again into sixty units. Thus, as Neugebauer notes, "our present division of the day into 24 hours of 60 minutes each is the result of a Hellenistic modification of an Egyptian practice combined with Babylonian numerical procedures."

Other legacies from ancient Egyptian astronomy are less obvious. Aside from the circumpolar constellations, and particularly the Big Dipper (called the Leg of the Ox by the Egyptians), there seems to be little relationship between the Egyptian view of the heavens and the familiar star patterns identified in today's zodiac. The thirty-six decans, or star groups, used for time-keeping were quite different from those used by either the Greeks or the Babylonians. Indeed, the Egyptians may not have been very much concerned with positional astronomy, since they had little practical use for the stars. As a sedentary people living totally enclosed in a narrow valley with the cardinal points clearly marked by the south–north flow of the Nile, they had little need for star guiders in navigation. The most famous Egyptian star map is the round zodiac of Dendera, but it dates from almost the Christian era (100 B.C.) and the temple shows a strong Greek influence, if not origin. The ceiling painting has a sky chart showing the Greek constellations of that period superimposed on the supposed Egyptian star groups. While some researchers consider this an "astronomical Rosetta stone," it is unclear how accurately the Greeks interpreted the ancient Egyptian sky symbols.

94

On the whole, Egyptian astronomy is disappointing: perhaps because we expect so much from it. As E. C. Krupp has pointed out in *In Search of Ancient Astronomies:*

> There was no systematic and comprehensive observation of the sun, moon, planets and stars, or at least no evidence of such observation remains. Nor is there any distinctly Egyptian technical knowledge for astronomical phenomena. Despite these shortcomings, Egyptian astronomers must have existed, for their contributions to modern civilization—the tropical calendar and the twenty-four-hour day—continue to remind us of the ideas they developed more than five thousand years ago.

The interweaving of astronomical imagery into the myths, legends, and poetry of the Egyptians also underlines the close interrelationship they felt existed between the terrestrial and celestial worlds. Even the simple geographical distribution of tombs and temples in ancient Egypt had a relationship to heaven and earth. Temples, symbolizing rebirth and life and associated with the sunrise, were located on the east bank of the Nile; tombs, symbolizing death and afterlife and associated with the sunset, were located on the west bank.

Other solar connections abound in Egyptian mythology and religion. Chief among the pantheon of deities was Amen-Ra, the sun god, whose being was manifested in the solar disk. (During his brief rule, the heretical pharaoh Akhnaton even attempted to establish the actual sun, called Aton, as the one true god.) The obelisks capped with golden pyramidions, or *ben-bens*, were monuments erected to Amen-Ra, and presumably the bright tips were intended to reflect his holy light. Some experts suggest even the great pyramids themselves were erected as massive tributes to Amen-Ra, for the slanting sides symbolized the sun's rays reaching down from heaven to earth. (This image of wisdom, light, and goodness persists in modern times as seen in the "pyramid and shining eye" on the reverse of the United States dollar bill.) More interesting, however, are those monuments that seem to have been deliberately constructed to catch the sun on astronomically significant days.

Encouraged by the discovery of several solstice orientations in Grecian temples, Sir Norman Lockyer traveled to Egypt to seek out other monuments that incorporated a long passage with the axis aligned so that sunlight, or light from another celestial object, might beam down the hallway and into a darkened inner sanctuary on a specific day. If the path was long enough, five hundred yards or more, then only the most precisely aligned beam would reach the sanctuary for a few moments on the proper day. According to Lockyer's calculations, the Egyptian astronomers may have used this method to determine the tropical year to an accuracy of about one minute.

In *The Dawn of Astronomy*, Lockyer cited six such alignments related to the summer solstice. Three were at Karnak in ancient Thebes (near Luxor). One of these was the alignment of the Great Temple of Amen-Ra, which seemed to point to the summer solstice sunset in the year 3700 B.C. Another, smaller, temple at Karnak oriented with the summer solstice sunset, and a third aligned with the winter solstice sunrise. However, most of the book was taken up with temple alignments with specific stars, taking into account the precession of the stars since the epoch in which Lockyer assumed the monuments were constructed.

Gerald Hawkins, who knew Lockyer's star alignments had been made invalid by revisions in the assumed dates, attempted to confirm the solstice alignments in Egypt. With Smithsonian Institution support, Hawkins surveyed Karnak and several other temples. Hawkins confirmed Lockyer's contention that the smaller temple of Ra-Hor-Ahkty southeast of the Great Temple complex was aligned to the winter solstice sunrise. But his results showed some surprises. For example, he found that the Great Temple of Amen-Ra itself seemed to align to the east and with the winter solstice sunrise rather than with the summer solstice sunset as Lockyer had thought. Certainly the eastern horizon had more significance for the ancient Egyptians; the heliacal rising of Sirius and the decans were all observed in this direction. Most texts referred to the morning stars, rather than to any evening phenomena. Moreover, the rising slopes of the Theban hills blocked any observation to the southwest of the winter sunset point as seen down Amen-

Ra's long and narrow passage. While the temple's axis line fitted an observation of the sun in reverse direction to the southeast precisely, the view down the axis in this direction was blocked by the Hall of Festivals. Hawkins thus looked for other places within the immense and sprawling complex where the view of the winter sunrise might have been preserved. At the far southeast corner of the temple, Hawkins was shown a small room on the roof, open to the sky, and reached only by a narrow staircase. This room, also dedicated to Ra-Hor-Ahkty, or the "brilliant horizon rising sun," had an altar from which one could look through a window opening onto the southeast (see Figures 28 and 29). Although a mud-brick protective wall built some thousand years after the temple now blocks the view, the horizon and the midwinter sunrise could easily have been seen in ancient times. Incidentally, the observing room stands above the long northern wall of the main temple. On this wall is a mural showing the pharaoh "laying out the cord," a ritualistic alignment that preceded each new building project.

Across the Nile from the Karnak complex Hawkins found, as Lockyer had suspected but never confirmed, that the Colossi of Memnon faced the winter solstice sunrise. Apparently these two statues once guarded the entrance to a temple aligned in the same direction. The temple is gone now and the colossi sit alone and in crumbly disrepair. According to local legend, when the colossi were in even worse shape and their vital parts lay scattered about their feet, the statues emitted a low hum, or whistle, when first struck by the rays of the rising sun. A well-meaning government archaeologist partially restored the colossi and they have never sung again.

Traveling farther south to the former Nubian lands below Aswan, Hawkins investigated the magnificent ruins of Abu Simbel, one of the world's finest monolithic sculptures. This temple, like so many others in Egypt, was constructed in honor of that supreme egotist Ramses II, specifically to mark the thirtieth jubilee of his long rule, around 1260 B.C. (see Figure 30).

The temple faces east so that the great statues of Ramses, as well as the effigy of Ra-Hor-Akhty that stands over the temple door, greet the dawn (see Figure 31). Indeed, each day as the sun rises

Figure 28. *A steep staircase leads to the small roof-top temple at Karnak, known as the High Room of the Sun.* Photo courtesy Smithsonian Institution and Gerald Hawkins

Figure 29. *The High Room of the Sun at Karnak.* Photo courtesy Smithsonian Astrophysical Observatory and Gerald Hawkins

Figure 30. *The temple complex at Abu Simbel. Behind this natural façade is an artificial mountain, for the entire structure was moved in the 1960s to its present location—and higher elevation—to escape the rising waters of Lake Nasser.*

behind the cliffs across the Nile, the light gradually slides down the huge stone bodies carved from the mountain, illuminating the statues almost lovingly inch by inch. On two days out of the year— October 18 and February 22—the angle of the sun is such that its rays penetrate into the rock façade and pour down a 60-meter-long passage to an inner sanctuary where they fall directly on the statue of Ramses II, who stands with Amon and Ptah at his right and Ra-Hor-Akhty at his left (see Figure 32). Interestingly, that October day when Ramses II is momentarily bathed in light by the rising sun corresponds within a day or two of the beginning of the Egyptian civil year at the time of Ramses' jubilee celebration. As Hawkins noted in a Smithsonian Institution report: "This simple result radically alters the interpretation of the temple—it was carefully site-selected and astronomically planned ahead of time so that the flash of sunlight would alight on the pharaoh god effigies, bringing life and rebirth to Ramses and starting a process of deification."

Could it be that this massive structure was carefully planned and then carved from solid rock merely to capture a brief burst of light? Aside from the evidence of other Egyptologists that suggests Ramses II became an "equal to the gods" sometime around his

Figure 31. *Two of the monolithic statues of Ramses II that flank the entrance of the main temple at Abu Simbel.*

thirtieth year, Hawkins found additional proof that Abu Simbel's architects were fully aware of the sun's swing along the eastern horizon. He discovered that the small antechamber, or side temple, which is set at an angle to the main temple and which appears mismatched with the rest of the complex, is aligned to the rising sun on the day of the winter solstice (see Figure 33).

Figure 32. *View of the temple entrance from the inner sanctuary at Abu Simbel. On two days of the year, the rays of the rising sun shine directly down this passageway to fall on a statue of Ramses II.*

Figure 33. *The secondary temple at Abu Simbel, found by Hawkins to be aligned with the rising sun on the winter solstice.*

Abu Simbel is now easily reached by plane from Cairo or Luxor, and most tourists fly in and out in a single day. Many are surprised to learn that the giant statues and all the interior stonework of the original temple have been moved from a location many yards below on the river bank. An international team of restoration engineers supported by UNESCO saved the temple from the waters of Lake Nasser, rising behind the high dam at Aswan. To the credit of this team, the ancient astronomical alignments of Abu Simbel have been perfectly preserved.

102

The most famous monuments of Egypt are also the most abused in the pursuit of astronomical connections. Of course the pyramids, by their very size and dimensions, beg for some explanation, astronomical or other. The Great Pyramid of Cheops, for example, covers 13.1 acres. Its four faces, slanted at an angle of 51°52', merge at a point 481.4 feet above the ground. Originally the pyramid was covered with a white granite facing, but this stone was long ago quarried away for the buildings of Cairo. The original gold-coated capstone is gone also; otherwise, the pyramid would be another 31 feet high. The pyramid is constructed of more than 2,300,000 stones, averaging 2.5 tons each, with some weighing as much as 15 tons. All this stone, covering so large an area, was laid with only simple surveying equipment; yet the corners are nearly perfect 90-degree angles and the greatest discrepancy between the length of the sides is only 7.8 inches. (The actual measurements are: north face, 755.43 feet; south, 756.08; east, 755.88; and west, 755.77.) Moreover, the four base lines are aligned with remarkable precision to the cardinal points of the compass, with the worst error, on the east face, only 5'30" west of north. (The errors in the other baselines: north, 2'28" south of west; south, 1'57" south of west; and west, 2'30" west of north.) The larger discrepancy in the north–south alignment may actually be the result of the slight shift in the earth's crust over the last four thousand years due to continental drift.

The interior of the Great Pyramid is a maze of chambers and passages. The so-called descending corridor begins at an opening on the north face about 55 feet above the ground and extends at a fairly steep angle some 345 feet into the pyramid stone and on into the bedrock below. The last 29 feet of this corridor are horizontal and end in a subterranean chamber. The ascending corridor rises from the descending corridor at a point about 60 feet inside the north face entrance. It rises for 129 feet before a side passage forks off horizontally toward the midpoint of the pyramid and the "Queen's Chamber." The ascending corridor continues upward to become the "Grand Gallery," a corbeled vault 28 feet high and 153 feet long. This vault ends in the "King's Chamber": a large rectangular room, 34 feet long, 17 feet wide, and 19 feet high. The room's main axis runs east–west and a series of smaller rooms, or "apartments," is located above it, formed by slabs of limestone arranged in layers almost like a series of false ceilings. A series of elaborate rock plugs once blocked the entrance to the King's Chamber. And from both the north and south walls of the chamber, narrow air shafts extend to the corresponding outer faces of the pyramid. (At one point, a second set of air shafts was begun to be dug through solid rock to the Queen's Chamber. They were never completed.) A crude vertical shaft connecting the ascending and descending corridors and then extending down into the bedrock may have served as an exit route for workmen during construction; or, as its popular name implies, it may have served as a "well."

One of the many mysteries surrounding the Great Pyramid is why no artifacts were found inside it. There is no evidence that Caliph Abdullah Al Ma'Mun, a ninth-century Moslem ruler of Egypt, found anything when he chopped his way through rock to reach the inner chambers (Ma'Mun's forced passage was cut into the north face about twenty feet below the original entrance. His men then connected with the ascending passage and, after cutting through the granite plugs, continued on to the inner sanctum.) Only an empty lidless sarcophagus was found in the King's Chamber and nothing in the other two vaults. (Most Egyptologists suspect that the so-called Queen's Chamber and the unfinished subterranean sanctuary in the bedrock were merely decoys to confuse looters.) The body of Cheops was gone long before Ma'Mun's men wormed their way into his tomb, if indeed it had ever been buried there.

The lack of mummies in either the pyramid of Cheops or of Khefren naturally has led to much speculation that the pyramids were really intended to serve a purpose other than as tombs. Among the suggestions: giant calculators, storehouses, granaries, sundials, and mnemonic devices for cataloging the ancient mysteries of geometry, algebra, calculus, and practical geography. And, of course, astronomical observatories.

Moses B. Cotsworth, a nineteenth-century Englishman obsessed with simplifying the Gregorian calendar, perceived the Great Pyramid as a giant device for predicting the length of the tropical year. According to Cotsworth, the pyramid worked much like a huge gnomon, with the shadow cast by the pyramid calibrated against inscribed lines on the surrounding paved surface to mark the solstices and equinoxes. He believed the pyramid cast its shortest shadow at the time of the two equinoxes, and, following the spring equinox, the noon shadow slowly retreated up the north face and eventually disappeared only to reappear and advance down the slope to reach the pavement at the autumnal equinox. In fact the Great Pyramid's minimum shadows occur some two weeks after the autumnal equinox.

The English astronomer and science popularizer Richard Procter, in his book *The Great Pyramid: Observatory, Tomb, and Temple* (1883), suggested that the pyramid was originally truncated just above the Grand Gallery. The gallery itself, then open to the sky, was used as a meridian transit sight. The passage of stars across the slit of the gallery could be very carefully timed, and the field of view could be widened or narrowed by the observer moving up or down the sloping passage. Presumably, once the sky had been charted and the stars timed with sufficient precision, the Grand Gallery's observing slit was bricked over, the pyramid completed, and the astronomical secrets locked inside forever.

The most influential of the nineteenth-century pyramidologists was Charles Piazzi Smyth, Astronomer Royal of Scotland. His misguided theories about the astronomical uses of the Great Pyramid, as well as the significance of the dimensions of that structure, still haunt—and mislead—the unwary. In 1864, Smyth published *Our Heritage in the Great Pyramid,* in which he described the "pyramid inch" (approximately one thousandth larger than the standard British inch) as one 500-millionth of the earth's polar diameter. Smyth claimed to have derived this "inch" from his own precise measurements of the pyramid (in truth, his surveying techniques were grossly inaccurate) and claimed this was the standard unit of measurement used by the pharaohs in the construction of not only the Great Pyramid but many other structures of the ancient kingdoms. Moreover, said Smyth, most of the physical laws of science, gained through centuries of study, research, and observational luck, could be found recorded in the dimensions of the pyramid. For example, Smyth believed that the perimeter of the pyramid base measured 36,524.2 pyramid inches, or almost exactly 100 times the length of the tropical year. Other measurements in the pyramid, including the length and breadth of the interior passages, stood for a period in world history that could be traced back to the construction date of the pyramid, which Smyth erroneously stated as 2170 B.C.

Although Piazzi Smyth is the best known of the pyramidologists, he merely elaborated on ideas first enunciated by the eccentric London publisher John Taylor, who believed that all the mathematical laws vital to mankind's salvation could be found within the pyramid's stones. The most important of these "truths" was the value of pi (π), which Taylor showed could be obtained by dividing the pyramid's height into twice the length of its base. And so it can be; for, as Kurt Mendelssohn has noted, "a pyramid with an angle of elevation of 52° has the unique geometrical property that its height stands in the same ratio to its circumference as the radius to the circumference of a circle, or $1/2\pi$." But as Mendelssohn also points out, the Egyptians probably "measured long horizontal distances by counting revolutions of a rolling drum. In this way they would have arrived at the transcendental number, 3.141 . . . without trying and without knowing."*

Other Taylor disciples regarded the Great Pyramid as a "Bible in stone"; and by the late nineteenth century, a veritable army of

---

* "The Egyptians used as a height measure the royal cubit . . . a length of 52 cm. Since ropes of palm fibre tend to stretch, a much more accurate way of measuring long horizontal distances as, for instance, the base of a large pyramid, was required. One such method is to roll a drum and count the number of revolutions. The royal cubit, already used for height measurement, would immediately suggest itself as the standard diameter of the drum, and one revolution . . . corresponds to the circumference of the drum. . . . It appears that using this system of measurement, the Egyptian architects never did anything more sophisticated than to build the pyramids according to the simple gradients of 4 : 1 and 3 : 1. Taking the former first, the height of the pyramid will be 4 × n cubits, where n is the number chosen to determine its size. The horizontal distance from the center of the building to its side will then have to be 1 × n rolled cubits or $n\pi$ cubits. Since this distance is half the side length of the pyramid, the latter's circumference comes to 8 × $n\pi$ cubits. Therefore the ratio of height is 4 × n/8 × n cubits, or by dividing this fraction by 4 × n cubits, simply $1/2\pi$."

—KURT MENDELSSOHN
The Riddle of the Pyramids

eccentrics, romantics, visionaries, and frauds were swarming over its stone tiers with tape measures and copybooks, checking, measuring, and discovering everything from the chronology of the Old Testament to prophecies of a new world order. Sometimes they even felt forced to shape the evidence to fit the theory. Sir Flinders Petrie tells of finding the proponent of one theory that depended on precise dimensions surreptitiously chipping away a corner of the pyramid base to bring the reality closer in line with his calculations.

Had pyramidologists such as Smyth restricted themselves to the numbers game at Giza, little harm would have been done. Unfortunately, Smyth was also an astronomer and he had no doubt that the Great Pyramid must also have been an ancient observatory. To Smyth it seemed obvious that those same passages and angles that mapped the course of human history must have also once mapped the course of the stars. (With good reason, perhaps, generations of archaeologists thereafter have remained leery of astronomers poking about their ruins.) Smyth, an astronomer of some standing, touched off a near-fanatical search for other astronomical alignments in the Great Pyramid that continues today. Naturally most investigators sought some sort of alignment with Sirius because that star had such a central importance in the life of ancient Egypt. The descending corridor of the pyramid has often been called the Sirius shaft, even though it points north and could never have been used to observe the star.

Smyth himself was convinced that the descending corridor, with its angle of 26°31′.4 to the horizontal, had been deliberately aligned with the star Thuban—that is, Alpha Draconis. According to his calculations, during the epoch of the pyramid's construction, that star was close enough to the celestial pole for its light to shine down the corridor once each night as it reached lower culmination, the lowest point in its circumpolar path. Unfortunately, Smyth believed the pyramid to have been built in 2170 B.C., when, in fact, it was constructed some five hundred to six hundred years earlier, and Alpha Draconis would not have worked.

Smyth also thought the pyramid was somehow associated with the Pleiades. According to his theory, when Alpha Draconis reached lower culmination at midnight, the Pleiades transited to the south. Obviously, the Pleiades could not be seen at this moment from the interior of the pyramid, but Smyth felt the relationship between Alpha Draconis and the Pleiades was celebrated by the seven courses that rose in corbeled fashion over the Grand Gallery. He also claimed the ascending passage was aligned to the Pleiades, but his calculation was at least 30 degrees off the mark.

Although Smyth's alignment of Alpha Draconis with the descending passage proved erroneous, David Allen recently has suggested that there might have been an alignment with Gamma Crucis. This crosslike grouping of stars has a distinct resemblance to the ankh, a tau cross with a loop at the top used as the Egyptian symbol of life. The gamma, or northernmost star, in the group would have formed the top of the ankh symbol and the group could have been aligned properly in about 2400 B.C. This date is considerably later than the construction, however; besides, there is no mention in Egyptian records of this star cluster. Even Allen, who tries hard to find some astronomical alignment, must admit the difficulty in his search. "The inner structure of the pyramid is not certain," he writes. "But the similarity of the inclination of the passages to half the angle that the faces make with the ground suggests that mechanics rather than astronomy guided the designs."

The ventilation shafts, too, have long been described as sighting tubes. In reality they both make short horizontal turns just before entering the King's Chamber, and sighting up them would have been impossible. However, Virginia Trimble and Alexander Badawy have found possible astronomical alignments for both shafts. The northern shaft, which exits the King's Chamber at an angle of 31 degrees, points toward the upper culmination of Alpha Draconis. The ventilation shaft on the south face, angling out at 44°5' from the horizontal, aligns with the star Alnilam in Orion.

But why alignments at all? After construction, the walls would have been sealed so that any sighting up the shafts would have been impossible and the pyramid would have ceased to function as an observatory. Those who proposed an alignment with Alpha Draconis usually claimed it could have helped maintain the north–south orientation of the pyramid during construction. But, in fact, Alpha Draconis wasn't close enough to the celestial pole to achieve the accuracy found in the pyramid's actual layout. Moreover, such an elaborate (and still inaccurate) device was not really needed. "A likelier method," writes L. Sprague de Camp in *The Ancient Engineers*,

> . . . is to build an artificial horizon—that is, a circular wall, high enough so that a person seated in the center cannot see any earthly objects over the top of the wall. The seated observer, with his head at the center of the circle, watches a star rise and directs another surveyor to mark the place where the star appeared. When the star sets, he causes another mark to be made. By lowering a plumb bob from the marks on the wall, the places at the foot of the wall, inside, and directly below the marks are found. Lines are drawn to the center of the circle. By bisecting the angle between these lines with cords and markings, true north is found.

Most likely, the airshaft alignments were never intended for observations, although observation might have been used to establish them. Trimble and Badawy suggest the shafts were linked to the "stellar destiny" of the pharaoh. After his death, he would join the other "imperishable stars" of the firmament, those circumpolar objects which neither rise nor set; in short, the stars of immortality.

It is this mixture of astronomy and religion, this commingling of myth and reality, and this application of advanced observing, engineering, and surveying to the purposes of fantasy that so frustrates and fascinates the student of Egyptian life and science. In Egyptian astronomy, science was totally subjugated to religion, and any practical advantages of observational ability (with the exceptions of time-keeping and surveying) were given over to the impractical and obsessive preparations for the afterlife.

But so, too, were all other aspects of Egyptian society. Indeed, despite its long history, its many monuments, and its beautiful art, Egyptian civilization culminated early. Within a span of only a few centuries out of several millennia—including the brief period of

pyramid building, which Mendelssohn has called the "first application of large-scale technology"—Egyptian intellectual and technical
achievement reached a peak, and thereafter stagnated. Thus, the
archaeoastronomer can find scores of examples of astronomical orientations in tombs and temples, but any evidence for the
emergence of mathematical science is slim. In Egypt, as in so
many other ancient cultures, the real purpose of astronomy seems
to have been astrology.

Oddly enough, although the Babylonians pursued astronomy for
the very same reasons, the astronomy of Mesopotamia is somehow
considered more scientific. Perhaps this is because historians traditionally have considered that "scientific" or mathematical astronomy requires not only the recognition and grouping of stars
and recording of cycles, but also the attempt to predict recurring
astronomical events, such as eclipses. The earliest Babylonian
priest-astronomers certainly attempted—often quite successfully—
such predictions.

A large, but not always clear, record of Babylonian astronomy
comes to us on baked clay tablets inscribed with cuneiform characters. No observatories or astronomical instruments exist; nor do
the tablets give any clear idea of how the observations were made.
(This is in marked contrast with the state of archaeoastronomy in
northwest Europe, where great stone observatories survive, without any clear record of their observations—or so some proponents
suggest.)

Babylonian astronomy is generally divided into two periods:
from prehistory to the destruction of Nineveh (607 B.C.), and from
the founding of the Neo-Babylonian Empire to the time of Christ.
Astronomy in the first period was comparable to that of the Egyptians in the same period: a mixture of practical observations and
socio-religious applications. In the second period, astronomy often
was studied for its own sake, although usually still for the same
end: to make astrological predictions.

At first the Babylonian calendar was based on the return of the
"apparent new moon," the first day on which the crescent moon,
having disappeared previously in the glow of the rising sun, reap-
peared in the western sky after sunset. Although this method es-

tablished the lunar month, the Babylonians somehow needed to match the lunar observations with the observed solar year. Like the Egyptians, the Babylonians soon discovered that the length of the tropical year could be found with fair precision by measuring the heliacal rising of a certain star. In practice, however, they attempted to find an appropriate star for each month, and if this rising was observed at the proper time then the calendar was in order. If the observation was out of synch with the civil calendar, another month was added to bring the solar and lunar years in order. Because not all stars were bright enough to serve as heliacal pointers, or because poor weather might interfere with some observations, the Babylonian luni-solar calendar apparently had a number of irregularities, especially in those years with the intercalated months.

Further divisions of time included some sort of week. The Babylonians considered each of the seven-day intervals (phases) of the moon to be distinct periods, and some were days of ill omen when a variety of activities were prohibited. The Babylonian day began with the instant of sunrise. Subsequent hours of the day were divided into twelve periods, called the *kaspu*. The path of the sun through the ecliptic during the year was also divided into *kaspus*— that is, the distance traveled in one day—for a total of 360 per year. This in turn led to the division of the celestial sphere and the circle into 360 degrees and marked the origin of the sexagesimal system, a legacy still with modern science.

The Babylonian astronomers—or more precisely, the Chaldeans, who were a part of the empire during the reign of Alexander the Great—also discovered a technique for predicting eclipses by finding that a definite interval existed between similar events. This was no small task, since lunar motion is quite complex. As we have seen, if the moon's orbit were in exactly the same plane around the earth as that of the earth's orbit about the sun, there would be both a solar and lunar eclipse each month, because the moon and earth would twice be perfectly aligned in relation to the sun and the shadow from one would fall on the other. But because the moon's orbit is tipped by an angle of 5 degrees, it usually passes either below or above the earth-sun line. The moon's orbit does pass through the earth's orbit twice a month; the points where it crosses are called the nodes. When the moon passes through one of these nodes and the line of the nodes—an imaginary line drawn through both nodes and the earth—is pointing toward the sun, an eclipse will occur.

Without any external forces acting on the orbit of the moon, this alignment of earth, moon, and sun would automatically occur at least twice a year on the exact same days. The moon's orbit is not fixed, however, and shifts slowly in respect to the stars while still keeping its 5-degree tilt. This precession of the nodes takes 18.61 years to complete a full rotation, so the line of the nodes realigns itself with the sun twice every 346.6 days rather than twice every 365.25 days. This is the so-called eclipse year. Over the short term, an observer probably would not notice any cycle to the eclipses. But with observations over a longer period, a repetition of an eclipse cycle might be noticed. Since the moon comes into a near alignment with sun and earth every 29.5 days, an eclipse will occur over the same general area of the earth whenever these two cycles and the tropical-year cycle all coincide. These cycles come into phase every 6,585 days, or 18 years and 11 days. This period is called the saros cycle. A persistent observer and accurate record-keeper might eventually realize that if a certain sequence of eclipses took place in a particular year, the same sequence would be repeated eighteen years later.

Since the Babylonians had been systematically making astronomical observations from about the eighth century B.C., their records alone would have allowed them to predict eclipses. Clay tablets left by the Chaldeans indicate they realized that every lunar and solar eclipse was part of a larger set of eclipses taking place at regular intervals. Each sequence generally included five or six eclipses, followed by a period of eighteen years when no eclipse in that series took place. Naturally, several periods overlapped, so there was always an eclipse of the moon in every year; yet the Chaldeans seemed capable of predicting the recurrence of specific eclipses far in advance. One of the earliest records of such a prediction was the carefully observed solar eclipse on March 10, 721 B.C.

The Mesopotamians probably had their best luck predicting

lunar eclipses, simply because the earth's shadow is so large that nearly half the globe will see any lunar eclipse at one time. By contrast, the shadow cast on earth by the moon is so small that a total solar eclipse will be seen within only a very narrow path.

Not everyone agrees that the Babylonians discovered the saros cycle. Giorgio Abetti points out that "even with a long set of observations, missing a few, through bad weather or other reasons, would be enough to disrupt the sequence of events." Perhaps, he suggests, they had some less complicated method, based on other periodic cycles of shorter intervals.

In the end, of course, the results of this eclipse prediction were essentially social rather than scientific, for they were intended primarily to ward off evil spirits or to signal danger periods for imperial decisions. But this is of little matter to archaeoastronomers. More important than the ends were the means of this astronomy. By divining the futures of kings from the stars, the Babylonians and Egyptians also designed the future of modern science. The precision observations, the accurate recording, and the mathematical expertise necessary to calculate the motions of stars, planets, sun, and moon formed the basis for the technical and intellectual flowering in the Hellenic period that followed. In a sense, our view of the heavens—and all the physical world—today has been shaped by the ancient visions of those first priest-astronomers.

Old Science in the New World

> *"Ancient records tell us that our predecessors scaled great astronomical heights. . . . In the Old World of the Mediterranean, they created most of the astronomy with which we as historians are familiar. On the American continent, other races, entirely separate from those of the Old World, also created a sophisticated system, an astronomy of equal brilliance."*
>
> —Anthony Aveni

Getting to Ihuatzio is not an easy task. A four-wheel-drive vehicle is best; I drove a Volkswagen bug. The first twenty miles are relatively simple; you drive west from Morelia on Mexico 15 and then turn south on the road to Pátzcuaro, an old colonial city famous for its Christmas festival and *pescado blanco,* a local white fish taken from the lake of the same name by Indian fishermen using elaborate "butterfly" nets of pre-Hispanic design. At the marker for the melodious-sounding village of Cucuchuchu, you turn onto a pothole-pitted dirt road that leads to the lake shore. In the farming settlement of Ihuatzio there are no signs marking the nearby ruins, so, primarily on faith, one follows vague directions to turn right at the village water tank onto a narrow, rocky cattle track through fertile fields where *campesinos* till the land with ox-drawn wooden plows. During the summer rainy season, when I unwisely chose to visit the site, the trail is muddy, slick, and in some parts an almost continuous stream studded with rocks. The ruins themselves are so unspectacular—no more than smaller, rounded versions of the surrounding volcanic hills—one could easily drive past them except that the trail dead-ends in a grassy cattle pasture about two hundred yards from the crumbling stone structures.

Although the reconstructed ruins of nearby Tzintzuntzán (once the administrative capital of the Tarascan empire) are better known and more extensive, Ihuatzio may have been more central to the cultural, religious, and scientific life of these exciting, artistic people, contemporaries and competitors of the Aztecs. Little excavation or reconstruction has been done at Ihuatzio. Three circular pyramids standing in a line parallel to the lake shore remain overgrown with brush and weeds. Neat rows of maize grow in what was once the large open plaza on their east side. Another one hundred yards north, again perfectly aligned with the lake shore, are two other structures, truncated pyramids with steeply rising stepped sides. Actually, these buildings are not true pyramids, for they are slightly elongated, with their bases more rectangular than square. They resemble somewhat the stone oratories built by the hermit priests of western Ireland. The similarity is even more striking because the "pyramids" at Ihuatzio are constructed of flat stone slabs laid in courses. However, unlike the oratories, which are hollow, or the Egyptian pyramids, which are solid stone, the Tarascan pyramids follow the Mesoamerican tradition of a stone facing laid over a solid core of hard-packed dirt and rock rubble.

The solitary watchman—usually the only person found at the ruins—serves as a one-man construction crew who, day after day, doggedly replaces fallen stones into their original positions. Ihuatzio has not yet received much scholarly attention. Most of the studies here, and particularly those concerning the astronomical orientations, have been done by Gabriel R. Muñoz, director of the new planetarium in Morelia, capital of Michoacán state. Using both ground surveys and aerial maps, Muñoz has found that the two pyramidal structures are perfectly aligned with the cardinal points—their long sides running north–south parallel to the lake and their short sides running east–west (see Figure 34). If one stands due east of the two structures and looks into the gap between them (they are separated at their bases by about ten yards) one can see the gigantic statue rising from the peak of Janitzio Island in Lake Pátzcuaro, now an important Roman Catholic

Figure 34. *View through the passageway between the two Tarascan pyramids at Ihuatzio, Michoacán, Mexico.*

shrine. Muñoz has found that the angle described by the sides of the pyramids, 45 degrees, corresponds to the angle of the noon sun above Ihuatzio on the day of the summer solstice. In other words, by standing at the east–west base of either pyramid at local noon on or around June 21 and sighting up the stepped side, an observer will see the sun stand directly above the truncated pyramid's flat top (see Figure 35).

In Tarascan cosmology, the universe was divided into three parts: the sky, which they called *Auandaro;* the earth, or *Echerendo;* and the world of the dead, or *Cumiechucuaro.* In turn, each part was further divided into five regions, four marking the cardinal points and the fifth at the intersection of the four directions. In each region dwelt a different god, represented by a specific color. From this division of the universe a Tarascan system of sacred numbers was derived: three symbolizing sky, earth, and nether world; four, the principal parts, or directions, of the world; and five, the totality of life and death. The most important deity in the pantheon of gods was Curicaheri, the sun god; and Ihuatzio was both his home and the center of this universe (see Figure 36).

Scattered throughout Mexico and other parts of Mesoamerica are literally thousands of ruins of the great civilizations that flourished for nearly a millennium before the arrival of the Spanish. Many of these sites remain unexcavated and, to a certain degree,

Figure 35. *Two truncated pyramids at Ihuatzio, Michoacán, Mexico. The slopes of the short sides point to the position of the midday sun on the summer solstice.*

Figure 36. *A Tarascan symbol of the universe. The snail-like design incorporates many of the Tarascan sacred numbers and concepts: the four cardinal directions; the three sides, each with five crenelations; and the five-band spiral. The in-turning spiral itself represents the sun god residing at the center of the universe.* Illustration courtesy Gabriel Muñoz

undeciphered. In the past two decades, as archaeology has expanded to include contributions by scientists from other fields, an increasing number of sites have been found to have either astronomical orientations or possible onetime uses as observatories. It is now recognized that many Mesoamerican cultures developed advanced techniques of celestial observation that led to precise timekeeping and calendar-making as well as to the discovery of basic celestial cycles. This early astronomy was as elegant and sophisticated as that of any Old World society, and, in some cases, much more accurate.

Ironically, the astronomical abilities of the ancient Americans were both long realized by archaeologists and long dismissed as secondary to the real business of understanding political history. Even Eric Thompson, one of the greatest Mesoamerican specialists, offhandedly dismissed the subject by noting that "the pathway to Mayan astronomy is strewn with the boulders of coincidence." Of course, it is possible that the early investigators didn't recognize this budding American science simply because it did not fit their concept of what science was. The goals of ancient astronomy were much different from those of modern science. "We are trying to understand our relationship with the natural world and control our environment," writes Anthony Aveni. "The Mayas hoped to predict the negative forces in the world so they could be avoided." But so did the Egyptians, the Babylonians, and the builders of the megalithic monuments in Britain. Archaeoastronomers feel the ultimate purpose of the observations makes little difference, for the New World astronomers were still practicing basic science. Given another millennium (or perhaps just another hundred years) untouched by European influences, New World "astrology" might have evolved into a system that attempted to understand nature rather than simply avoid it.

The first of the great Mesoamerican civilizations (at least, the largest and most successful after the Olmecs) was that of the people who inhabited Teotihuacán. Today the remains of this extraordinary culture serve as Mexico's best-known tourist attraction. The proximity of the ruins to Mexico City helps draw tens of thousands of visitors each year to see "the pyramids."

Of course, tourists see only a small portion of the glory that was once Teotihuacán. At its peak, Teotihuacán covered about eight square miles and housed as many as a quarter-million people. The rulers of Teotihuacán governed an empire covering an area that ranged from Central Mexico to the Guatemalan highlands. Trade—primarily in obsidian—stretched from the imperial city thousands of miles north into the American Southwest and Mississippi Valley and as far south as the coast of Peru. Teotihuacán's dominance over Mesoamerica lasted nearly a millennium, reaching its peak in about A.D. 650. Its influence and heritage would be felt in every subsequent Mesoamerican culture. Among Teotihuacán's most important contributions would be a knowledge of positional astronomy and the use of astronomical orientation for the precise arrangement of ceremonial centers and individual buildings.

Even the most casual visitor bused out from his hotel as part of a package tour senses an immediate connection between Teotihuacán and the sky. After all, the main pyramids are dedicated to the sun and the moon (see Figures 37 and 38). But the real astronomical connection may be more subtle. Archaeologists have found evidence that the great sprawling city was actually planned quite carefully; and once the plan was decided upon by the designers, all subsequent buildings in the complex strictly adhered to the blueprint. The city's gridlike street plan was apparently also adopted in many other urban areas that came under Teotihuacán's influence.

Oddly enough for such a careful layout, the axis of Teotihuacán itself is not the north–south line one would expect. Rather the axis is on a line running 17 degrees to the east of astronomical north. This characteristic skewing of the city's main axis appears again and again at the major ruins of Mexico. Tula, north of the modern Mexico City and the former capital of the Toltec nation which replaced Teotihuacán, follows the same offset pattern. Most Mayan cities in the Yucatan, which were influenced in turn by the Toltecs, show a skewing from the north–south line, ranging from as little as 7 degrees to as much as 23 degrees east of north. This effect can be seen most prominently at Chichén-Itzá, Dzibilchatún, and Mayapán. At the last site, the central ceremonial buildings are cardinally oriented, but all other buildings are offset to between 11 and 18 degrees east of north. What is the reason?

To some, the orientation suggests an astronomical alignment, but the actual connection remains unclear. However, recent investigations by Anthony Aveni of Colgate University and Horst Hartung of the University of Guadalajara have offered some tantalizing clues. Aveni suggests the orientation may be explained by two unusual figures carved in the stones of Teotihuacán. Actually, while one of the petroglyphs is located at the center of the city, the second is on the slope of a hill some three kilometers west of the city but within line of sight of the first carving. The petroglyphs

Figure 37. *The Avenue of the Dead at Teotihuacán, Mexico, with the Pyramid of the Sun on the left.* Photo courtesy Mexican Government Tourism Office

Figure 38. *The Pyramid of the Moon at Teotihuacán, Mexico.* Photo courtesy Mexican Government Tourism Office

are what Aveni calls "pecked crosses," small dots, or circles, chipped from the rock with a sharp instrument to form a "double circular pattern centered on a set of orthogonal axes."

A line drawn between the two pecked crosses, the one on the hill and the other in the city, aligns within seven minutes of arc of being perpendicular to the north–south axis of Teotihuacán. More important, if the line between the crosses is extended to the western horizon, it marks the place where the Pleiades star cluster set in 150 B.C. about the time when Teotihuacán was constructed. Interestingly, during this epoch the Pleiades also rose heliacally on the same day that the sun made its zenith passage, passing directly overhead at noon. (Ethnology and history support the notion that the Pleiades were important to Mesoamericans, particularly the Aztecs, who held special ceremonies to watch the zenith passage of these seven bright stars.)

The same pecked-cross symbol has been found in a variety of other locations throughout Mexico, suggesting that Teotihuacán's builders carried the symbol—and its purpose—with them in their colonizing ventures. The pecked cross is also similar to the "quartered circle" found throughout Mesoamerican art.

For example, a stone slab marked with two concentric circles of holes was found near the United States border by the Mexican poet and amateur archaeologist Alfredo Chavero in 1899. The circles were divided into quadrants by two diagonal lines; the outer circle had 25 holes per quadrant, the inner circle 20 per quadrant, and each axis had 20 holes, for a total of 260, excluding the central point.

This combination of 20 and 260 has an apparent connection with the pre-Columbian calendar. The Maya's ceremonial year had 260 days, and both the Mayas and Aztecs based their numbering system on the unit of 20 and multiples of 5, with special names given to 5, 10, 15, and the powers of 20. The Tarascans, too, as mentioned earlier in this chapter, subdivided their universe into five regions.

The axial arrangement of the cross had ten holes between the center and the first inner circle, four holes between the circles, and another four beyond. This 10, 4, 4 sequence suggests to some that the artist was representing an 18-month tropical year.

Aveni found a similar cross-and-circle device at Uaxactún, Guatemala, where it was pecked into the floor of a structure whose axis is offset from astronomical north by about 17.5 degrees. This axis seems to have been skewed intentionally so it would point toward Tikal, the largest and most spectacular of the classic Mayan jungle sites, some nineteen kilometers away (see Figure 42). Indeed, an observer standing on the pecked-cross symbol and looking south can see the Temple of the Jaguar at Tikal above the treetops.

Another nineteen similar symbols have been found by Aveni and his coworkers at Teotihuacán and elsewhere. They vary widely in both design and execution, but generally the pecked crosses are about a meter square and show the double circle. However, one

Figure 39. *Aztec ceramic stamp designs showing the pattern of the ancient game* patolli, *similar to the designs of the "pecked crosses."* Illustration from *Design Motifs of Ancient Mexico*, by Jorge Enciso. New York: Dover, 1953

symbol displays a double-square border, and one consists simply of the crosses. Eleven have been pecked into the floors of buildings; the rest have been found on rock outcroppings with unobstructed views of the horizon in several directions.

The "quartered circle" or "pecked cross" symbol may have had several interrelated functions in Mesoamerica, according to Aveni. For example, at Teotihuacán the axes of the pecked crosses seem to match the layout of the streets and buildings, and therefore may have been used by architects as benchmarks for following the grid system. By contrast, those symbols found outside of buildings on the distant hilltops or on stones remote from any settlements may have had an astronomical motive, perhaps providing alignment with some important celestial body on the horizon. At the same time, the numerology of the holes seen in the more complete and elaborate crosses suggests use as calendars. Finally, there were also several ancient games, such as the Aztecs' *patolli* (similar to Parcheesi), that used a cross-shaped arrangement of holes with pebbles or dried beans as markers and split reeds as dice (see Figure 39).

There is no reason the pecked-cross symbol could not have served all these purposes. Modern society insists on compartmentalizing life. Ancient peoples had no such prohibitions: All of life was intertwined. "The habit of distinguishing between dif-

Figure 40. *The Caracol at Chichén-Itzá.*

ferent intellectual phenomena and attempting to establish precise definitions of science (and art, technology, and religion) belongs particularly to Western civilization," says Ubiratan D'Ambrosio. "But if we study the old Peruvian *quipu* counting cord, for instance, we see clearly a scientific element side-by-side with a divine principle embodied in the Inca."

Nowhere is this mixture of the mystical and the mathematical, the integration of the divine and the practical more apparent than in Mayan astronomy.

The Mayas rose to power in Central America around A.D. 400 and remained a major force in the civilization there until the arrival of the Spanish eleven hundred years later. Their exact origin as a distinct people is unclear; however, like so many other Mesoamerican cultures, the Mayas probably descended from one of the many wild nomadic hunting groups that pushed south to settle as agriculturalists in the more hospitable tropical climes. It remains a mystery, however, why they chose the steamy jungles of Guatemala, Honduras, and Belize as the seat of their great kingdom. (Some experts suspect the presence of numerous natural limestone "water tanks" encouraged settlement here.)

Mayan civilization blossomed rapidly, borrowing much from the Olmecs who preceded them, and reached a peak in about A.D. 700, the Classic Period. Then, for some reason still not clearly understood, between A.D. 800 and 925 the Mayas abandoned their great cities in Guatemala, including the elaborate ceremonial center at Tikal, and migrated northeast to Mexico's Yucatan Peninsula. Most scholars suspect the Mayas were victims of ecological exhaustion; they simply overtaxed the fragile environment of their jungle home. The Mayas' greatest scientific and mathematical achievements were accomplished before this forced migration; but the same traditions were carried on in the so-called Post-Classic Period that continued until the arrival of the Spanish. In the Yucatan, however, the Mayas also came under the influence of the Toltecs. Many Toltec astronomical, architectural, and religious traditions, including ritual human sacrifice, the oddly skewed grid of cities, and the legend of Kukulcán-Quetzalcoatl, were incorporated into Mayan society.

Although archaeologists often describe the Mayas as "pre-literate," this is not strictly correct. We simply find it difficult to read their writing. The Mayas developed an elaborate hieroglyphic system, part syllabic and part rebus writing, using anthropomorphic and zoological forms and a color code. To date, the messages left by the Mayas are only partially understood. The problem of studying their early science is exacerbated by the paucity of information. For example, the carvings on the temple walls and the monumental stelae at most jungle sites carry some notes on Mayan life and thought, but the prime function of these "writings" was ceremonial.

The real "texts" of the Mayan scholars—and the "operator's manuals" of its technicians—were the codices, and only a pitiful few survived the Conquest. The codices were painted books made from pieces of deerskin or panels of processed tree bark and then sewn together in long strips to form folding screens of many pages. In a real sense, the codices were Mayan encyclopedias. They were compendiums of science and technology, including astronomical and mathematical knowledge as well as more practical instruction in agriculture and arts and crafts.

Unfortunately, in July 1562, at a well-attended religious ceremony in Mérida, Diego de Landa, the first bishop of the Yucatan, destroyed over a score of these codices by fire. He called the books the devil's work, and, by his act, wiped away vital clues to the mysteries of Mayan hieroglyphics. (Ironically, in his later years de Landa decided to record the life and culture of the Mayas. And it is from his classic history that we draw most of our knowledge of Mayan civilization in the years immediately preceding and following the Conquest.) Only a handful of the codices survived the religious zealots and somehow made their way into European museums. In short, what we know about Mayan science is based on an extremely small and incomplete record.

The Dresden Codex (named for its current home in that German city) provides the clearest evidence of Mayan mathematical genius, for it carries an elaborate and detailed calendar for observations of Venus and a ritual almanac for predicting eclipses. An eclipse was obviously of supreme importance, for that was the time when one of the Mayas' primary deities disappeared.

The lunar phase tables in the Dresden Codex, described in great detail by other authors, show repeated sequences of 177 days, representing 6 lunar-phase months (full moon to full moon), and 148 days, representing 5 lunar months.

If the Mayan astronomers had watched and recorded lunar eclipses carefully, they would have soon discovered that eclipses were most likely to coincide with every sixth full moon, or every 177 days. As has been mentioned before, however, eclipses cannot take place unless the earth and moon lie along a nearly straight line pointing toward the sun. This alignment occurs approximately every 173 days. Obviously, this true period of eclipse half-years, or 173 days, invariably creeps ahead of the full moons, and the Mayas must have known it, for they seem to have occasionally inserted a five-month period (148 days) into their calendar to make the two periods correspond.

The Dresden Codex and others make it abundantly clear that the Mayas considered time to be cyclic, but some cycles could extend over several millennia. For example, their "long count" of universal existence began August 13, 3114 B.C.; and the passage of time from then was counted much the way an automobile's odometer clicks off mileage on a long trip. Starting at that creation date, which Kenneth Brecher and Philip Morrison think corresponds to a solar zenith transit over the Mayan city of Copán in Honduras, time was clocked in various units, adding up regularly to 260-day years, each with either 18 or 20 months, but with both the dates and the day counts becoming cumulatively larger.

Within this apparently endless period, there were many shorter cycles, including the ritualistic 260-day year and the 584-day period of Venus (from first appearance as morning star to reappearance as morning star), as well as longer cycles measured in hundreds of years, when all celestial events repeated themselves—with the expectation that terrestrial events would also repeat themselves. (One imaginative theory for explaining the wholesale uprooting of the Classic Mayas claims the priest-astronomers may have so successfully calculated the "end of time" that the entire society had to "end" and "begin" again.)

Why this multiplicity of calendars? Most likely the Mayas needed a variety of time-keeping systems for a variety of social purposes. John Carlson suggests that much of the complex calendric information was simply designed to "provide a cosmic sanction for the rule of the religious and political elite." He has found that some stelae, those intricately carved stone pillars recording historic events, often linked the date of the then-current ruler's birth with the birthday of some ancestral deity in the past, making the two dates numerically related. By this means, says Carlson, a Maya chieftain could "provide a mandate for his rule by inventing a complex cosmology in the mythological past that justified his political control in the present."

Although the Mayas refined the practice to an art, many other societies maintain multiple (or at least dual) calendar systems. Catherine Callaghan, writing in *Current Anthropology*, has noted that most "ancient agriculturists employed a corrected solar calendar to regulate agricultural activity and this calendar functioned concurrently with the elite [ritual] calendar. The two calendars had different natures and functions. While the elite calendar was progressive and was employed to record history, the common calendar was cyclic and used to predict seasonal change."

Thus, while ritual observations were sometimes adjusted to fit religious dates or political purposes, the Mayas apparently also maintained an accurate, real-time calendar of astronomical events. Their calculation of the astronomical year, probably by observation of the sun's zenith passage, was precisely 365.242500 days, or only slightly less accurate than the modern sidereal measurement of 365.241298 days. The lunar tables of the Dresden Codex also show the Mayas measured the length of the actual lunar month as 29.52592 days, or within 7 minutes of the best modern measurement. Furthermore, the Mayas calculated the average period of the synodical revolution of Venus with an accumulated error of only 1 day in 6,000 years. Alas, this does not necessarily mean the Mayas understood the revolution of Venus around the sun. They were interested in the planet's appearance and disappearance in terms of time, not space.

Although much of the Mayan calendar could be determined through numerology, precise observations were still necessary and, indeed, essential for the more complicated bodies such as

Venus. The Venus table of the Dresden Codex dates from the Post-Classic period (twelfth or thirteenth century A.D.) in the northern Yucatan; and the most famous astronomical structure of the Mayan culture, the Caracol of Chichén-Itzá, also dates from that time.

The Caracol (the name means "snail" in Spanish and the conquistadors may have been referring to the spiral staircase that winds up the interior of the cylindrical building) even looks like a modern observatory: round with a domelike roof (see Figures 40 and 41). Although few major pre-Columbian buildings were round (the Tarascan pyramids at Tzintzantzún and Ihuatzio are notable examples), the shape of the Caracol may have been more aesthetic than scientific. Yet a careful inspection of the structure shows that its small apertures, or windows, are placed in an oddly unsymmetrical pattern. The offset positions of opposing windows provide sightlines that align well with the setting of Venus on the western horizon. (Although the windows framed a rather wide area on the horizon, moderate precision could be achieved by use of the "alternate jamb" method of observation. In this technique, an observer uses the inside edge of a door or window frame as a backsight and the opposing outside edge or jamb of the opening as a foresight. Measurements of position would be made as an astronomical object crossed the line of sight formed by these two marking points.) Perhaps Mayan astronomers watched through these narrow slots to mark the disappearance of the planet in the west so that they could calculate its reappearance eight days later in the east, when it rose heliacally ahead of the sun and returned as the morning star. Venus was associated with the god Kukulcán-Quetzalcoatl, who, according to legend, disappeared into the west with the promise to return someday out of the Eastern Sea. (The arrival of Cortez from the sea seemed to the Indian rulers a fulfillment of that prophecy.)

Other buildings throughout the Mayan world show similar alignments with astronomical objects, most notably the sun. The temple complex at Uaxactún seems designed as a giant solar observatory. By standing on the top of what is called pyramid VII and facing east at different times of the year, an observer can watch the sun rise over three different temples across an open plaza. Although located deep in a jungle, the temples are raised on artificial hills, or earthen platforms, so they poke above the surrounding trees. When the sun rose over the northernmost of the three temples, it marked the summer solstice on June 21. Sunrise

Figure 41. *The Caracol at Chichén-Itzá.*

over the southernmost building marked the opposite, the winter solstice on December 21. And sunrise over the central building marked the two equinoxes, vernal and autumnal, when the hours of daylight and nighttime were equal and the sun rose and set due east and west.

At Copán, in Honduras, the priest-astronomers constructed a pair of stelae separated by some seven kilometers on hills to either side of the main building complex. Aveni argues that a sightline connecting the two stelae also points to the horizon position where the sun sets on April 12, the date on which these people began their agricultural year—that is, when they began burning off brush for planting in preparation for the rainy season. Brecher and Morrison, however, think the sunset alignment marks the day of the

Figure 42. *The main pyramid at Tikal, Guatemala. The ruins of Tikal represent the remains of one of the great jungle cities constructed during the Mayas' classic period.*

solar zenith transit, August 13. By coincidence—or maybe it was design—Copán also lies at precisely the right latitude for the noontime sun to pass south of the zenith 260 days a year, while on the remaining 105 days it passes north of the zenith. This 260-day period corresponded with the ritual year of the Mesoamerican calendar, and Aveni suggests Copán was selected specifically as a major center because there the religious calendar would be in harmony with the astronomical calendar.

Several other examples of astronomically oriented buildings have been found in Mexico beyond the Yucatan. Monte Albán, near the modern city of Oaxaca and some two hundred miles south of Mexico City, was once the capital of the Zapotec empire, a contemporary of the Aztec. Descendants of these peoples still live in the hills around Oaxaca, following a pastoral life little changed from the pre-Conquest days. In fact, Monte Albán still holds a special place in their religious life, although the narrow, winding road to the mountain site is usually packed with the cars and buses of foreign visitors. In recent years, Monte Albán has become an increasingly popular spot for tourists, perhaps because of its impressive setting high above the Valley of Oaxaca and the persistence of folk life and folk art in the surrounding countryside (see Figures 43 and 44).

Even the most casual visitors note one strange aberration in the otherwise symmetrical layout of the ruins, which follows the classic pattern of a large open plaza surrounded by temples and pyramids. Although the other buildings are either square or rectangular, with their bases oriented to the cardinal points, building J in the southwest section of the plaza is sharply turned at an angle to the others and has a shape like an arrowhead. Since the first serious excavations of Monte Albán in the 1930s, building J has been described as some sort of "observatory." However, only within the last decade were any systematic studies of the structure made, again by Aveni, to determine its exact—if any—astronomical alignments.

Assuming that any observations of a celestial event occurring on the horizon would most likely have been made from a doorway or window at the front of the building, Aveni laid out a sightline along the main axis of the building. This line looked to the northeast horizon in a direction (47.5 degrees) that corresponded to the rise of

Figure 43. *The Monte Albán ruins are actually the product of three cultures. The first, most likely of Olmec origin (700 to 300* B.C.*), built the site's first temples and platforms that would serve as foundations for later structures. The second group arrived in about 300* B.C. *and is thought to have originated in Guatemala. The third group (*A.D. *500–1000) was the Zapotecs; most of the buildings remaining today date from their habitation. Building J, the "arrow-head" observatory, is in the right foreground.* Photo courtesy Mexican Government Tourism Office

Figure 44. *View of Monte Albán overlooking the Valley of Oaxaca.*

Figure 45. *According to the Codex Mendoza, this figure represents an Aztec astronomer-priest "watching the stars at night, in order to know the hour, this being his particular function."* Illustration by the author, after Nuttall

Figure 46. *The crossed-stick symbol, often incorporating an "eye" or "star" design between sticks, appears frequently in Mayan codices and has been interpreted as signifying an "astronomical observation." When the symbol appears in the doorway or window of a building or temple symbol, it has been interpreted as representing an "observing place" or "observatory." In the illustration at the top, a priest wearing the rain-god mask and carrying the emblem of the sun faces a seated woman, or priestess, on a platform in front of an "observatory." Both figures appear to be pointing at a star in the forked stick between them. The two symbols below show variations in the "observatory" design found in various Mayan records. Note that the temple on the left is surrounded by eyes, traditional star symbols.* Illustration by the author, after Nuttall

usually this type of structure—the large religious and political buildings created from stone and mortar—that most often survives, rather than the temporary and temporal homes of workers and peasants, it is not surprising that an unusually high percentage of Mesoamerican ruins seem to have some astronomical orientation. This is especially true considering the role of astronomy (or astrology) in the integrated spiritual-political life of the Mesoamerican cultures.

Still, the cynics who suspect archaeoastronomers can find astronomical alignments everywhere they look should note the example of the Nazca lines.

Over forty years ago, Professor Paul Kosok of Long Island University was working in the coastal desert south of Lima, Peru, near Nazca, mapping what he thought were shallow irrigation ditches— long straight paths in the topsoil bordered by thin lines of stones. The mapping of one wildly twisting and turning pathway revealed a pattern of unusual and unexpected design: a giant bird (see Figure 47). Subsequent mapping—most done later from the air—has revealed more than thirty figures plus an incredible network of overlapping lines and geometrical shapes, all created apparently by removing a dark top layer of stones to reveal lighter-hued sand below.

The south coast of Peru has a range of low hills running north and south just inland from the sea. Between these hills and the Andean foothills is a flat basin filled with erosion materials that have flowed down from the mountains over thousands of years. After winds blew off the dusty topsoil, they left a hard surface of pebbles and small boulders that eventually oxidized to a dark reddish brown. In effect, the desert became a "natural blackboard," for if the dark upper rocks are turned over, a lighter area of sandy soil is revealed below. If the dark stones are placed along the edge of a cleared area it produces an effect (when seen in aerial pictures, at least) akin to outlining a drawing in heavy crayon. Because rainfall in this part of Peru is almost negligible, oxidization of the cleared areas is extremely slow and the color differences remain today much as they were two thousand years ago.

When and by whom the lines were constructed is now fairly well established. A pre-Incan culture centered on the Nazca Valley thrived here in the period between 200 B.C. and A.D. 600. The other material remains of these people, particularly the woven fabrics for which they are justifiably famous, all show geometric patterns and animal representations strikingly similar to the desert drawings. The dates can be further refined by the radiocarbon analysis of wooden stakes found in the desert that were probably once used as sighting posts for laying out the lines. The dating places these stakes at approximately A.D. 525, or very late in the Early Intermediate Period of the Coastal Peruvian Culture.

It is less clear why the drawings were made. They obviously cannot be seen easily from the ground, unlike other great "blackboard figures" such as the chalk giants on the hillsides of Britain. The Pan-American Highway cuts through the desert at Nazca, ac-

Figure 47. *The figure of a "walking bird" with oversized claws is seen in this aerial view just below the image of a modern truck traveling along the Pan-American Highway.* Photo courtesy Smithsonian Astrophysical Observatory and Gerald Hawkins

tually bisecting many lines and even an animal or two. Until a few years ago, when a viewing tower and road sign were finally erected, most motorists whizzed right through the area in blissful ignorance of the patterns around them. Even the most determined researcher finds it difficult to spot the lines from ground level; the thinner lines can be seen with some difficulty by straddling one and looking down its length toward the horizon. However, the line nearly disappears if you step slightly to either side. Larger cleared areas are indiscernible to anyone standing in them (see Figure 48).

Yet, seen from the air the designs and shapes are quite intricate and detailed—and puzzling! For example, one drawing seems to illustrate a species of spider found only in the Peruvian jungles on

Figure 48. *This aerial view of the Nazca desert shows a huge flower and several extended lines and cleared "pathways." The brighter, curving sets of parallel lines are apparently recent tracks left by modern vehicles.* Photo courtesy Smithsonian Astrophysical Observatory and Gerald Hawkins

the *eastern* side of the Andes. The actual spider is only a half inch in diameter and, in one of the odder twists of nature, has its reproductive organ on its third leg. Usually this organ can only be seen with a magnifying glass. Here in the desert, the spider stretches over several hundred meters and its nearly invisible reproductive organ is shown in gigantic detail (see Figure 49).

Nor were the drawings and lines easy to prepare. William Isbell estimates the human energy expended clearing ground for some designs was comparable to that needed to construct the larger temples of Peru. This is especially true of the geometric patterns, which may be a half kilometer wide and a kilometer or more long, including some that run nearly eight kilometers.

The question of what purpose these lines served has never been satisfactorily answered. As Paul Kosok continued his survey of what he soon realized were deliberate markings of unknown significance, he noted that one line pointed to the spot on the horizon where the sun rose on the day of the summer solstice. He did not pursue this idea himself; but he left Peru convinced that the lines represented ancient astronomical sightlines. Maria Reiche, a German-trained math and astronomy student from Lima, took up Kosok's lead and has continued astronomical investigations at Nazca ever since. Today she describes the lines as "ceremonial walkways" that also have some calendric purpose. Other, more sensationalist writers have proposed a host of explanations over the years, ranging from "messages to the Gods" to "runways for interstellar space travelers." Always, however, the belief in an astronomical purpose persisted.

In the late 1960s, Gerald Hawkins followed his work at Stonehenge with expeditions to Nazca in an attempt to find the proposed astronomical alignments. After weeks of surveying the site, he then fed the directions of the lines into a computer to be matched with the coordinates of celestial objects for the proper epoch. He found only a few correlations. In his final report to the National Geographic Society and the Smithsonian Institution, he wrote:

> The ancient lines in the desert near Nazca show no preference for the directions of the sun, moon, planets, or brighter

Figure 49. *Aerial view of the lines in the desert near Nazca, Peru, showing the large figure of a spider.* Photo courtesy Smithsonian Astrophysical Observatory and Gerald Hawkins

stars. Nor do the lines show any deliberate alignments with a fixed but identifiable object in the sky, such as a nova or the center of some ancient pattern of stars. Thus, the pattern of lines as a whole cannot be explained as astronomical, nor are they calendric.

Hawkins backed up his computer findings with some plain astronomical good sense. The visibility at Nazca simply was not very good for observing, and particularly not for seeing objects along the horizon. The constant dust and haze, if conditions were similar at the time when the lines were drawn, would have severely interfered with watching the rise of any star along the horizon. Moreover, in the nighttime darkness the lines would have proven useless as markers unless they were somehow lit. No evidence for the use of lamps has been found at Nazca; besides, the fires themselves would have interfered with observations.

Hawkins did find some individual lines matched up with the significant horizon points for the rising and setting of the sun at the solstices and equinoxes, just as Kosok and Reiche had suggested. In fact, about twice as many alignments showed up than would be expected by pure chance. Hawkins preferred to view the lines as a whole monument, however; and, in this context, he could find no significant astronomical connection.

In effect, the Nazca example is used by Hawkins (and other archaeoastronomers) to show that every oddball and mysterious structure need not have an astronomical explanation; and, more importantly, not every random collection of lines and angles, no matter how abundant, will automatically produce some astronomical solution. The ancient astronomers apparently knew what they were doing, for there is little ambiguity about a true observatory.

Of course, it is possible that Nazca lines could have some other indirect and more subtle connection with the stars. Hawkins, like so many other visitors to Nazca, was disappointed not to have found a solution to the mystery. He therefore suggested that the lines could be somehow related to the belief in spirits pervasive throughout the Andean cultures. These spirits are thought to be gratified only by a gift or a sacrifice, and Hawkins theorized that the remarkable abundance of pottery sherds found at the Nazca sites may have been related to offerings placed at the ends of certain lines.

The Incas, who followed the Nazca people, also had a tradition linking "lines" with spiritual powers. Imaginary lines radiated in all directions from the central Temple of the Sun in Cuzco, the Inca capital. Thomas Zuidema has proposed that some of these lines were related to the seasons as indicated by the sun's changing position. Priests at the central temple may have charted the sun's motion by observing its position in relation to four stone towers placed on the crest of a hill west of the city. As a priest-astronomer watched the sun move south during August (the end of late winter in the Southern Hemisphere) and noted its arrival at the different towers, he could forecast the approach of spring and determine the proper planting times for crops at the different altitudes of the altiplano. Isbell further suggests that these imaginary lines extended beyond the horizon to serve as a gigantic annual calendar of rituals and ceremonies throughout the Inca empire. "The marked lines of the Nazca Valley might be related to the unmarked lines of Cuzco," he wrote in a *Scientific American* article. And perhaps "they contained certain symbolic information mapped on the ground for successive generations to observe, recognize, and memorize. The mapping of calendrical data on the ground, if it was combined with ritual observations, could have communicated not only information of agricultural significance but also other kinds of information useful to a complex but preliterate society faced with a pressing need to store knowledge acquired from generations of experience.

"The storage of inventories by means of the well-known Peruvian string-and-knot system of enumeration, the *quipu,* was evidently a practice adopted early in the prehistory of the Andes. Calendrical data, particularly when their use calls for the cross-check of actual astronomical observation, might be impractical or impossible to store by means of *quipus.*" Isbell thus suggests that information was symbolically coded and recorded "in the most durable medium available: the surface of the earth itself."

The interpretation of ancient astronomy in the Americas is made more difficult by the diversity of motives which spawned it. Agriculture, politics, calendrics, and religion—the sacred and the secular—became mixed. While modern astronomers pursue the goal of knowledge for its own sake, influenced by a Hellenic tradition of studying nature out of pure curiosity, the pre-Columbian astronomer—who was often also a high priest of the state religion and a part of the power elite—sought to coordinate the sometimes arbitrary events of the natural world with the human machinery needed to maintain the political order and perpetuate the ruling class. Whether this socio-politico-religious function of astronomy would have someday developed into a more naturalistic approach to science had it not been for the intervention of Europeans remains one of the great unanswered questions of history.

Amerindian
Astronomy

"It is apparent that many of us have operated under a form of chauvinism that has failed to appreciate the integral role of astronomy in the lives of early peoples everywhere. Compounding this has been a degree of twentieth-century parochialism that vastly underestimates the abilities and intelligence of prehistoric and non-industrialized men and women."

—Kendrick Frazier

The first reaction upon seeing Casa Grande is wonder. Why, you ask, would anyone have built this massive adobe structure here in the middle of a scorching desert? The sun beats down mercilessly on this ancient Hohokam site south of present-day Phoenix, Arizona. The few stands of palo verde and greasewood provide little shade. Except for a multitude of scrawny ground squirrels and some minuscule lizards traveling at what seems to be nearly the speed of light, all other creatures, including the perspiring tourists and the National Park Rangers who guide them around the ruins, move very slowly, almost as if the soles of their shoes had fused to the hard-packed sand.

The Casa Grande National Monument is located approximately twenty miles east of Interstate 10 and some forty miles south of Phoenix, Arizona. Although there is a modern town of the same name twenty-five miles away, the ruins are actually on the outskirts of Coolidge, a somnolent wide spot in the road where one can still find old-fashioned corner cafes and one-room general stores.

A structural steel canopy covers the "big house," protecting it from infrequent rains and providing some shelter from the sun for modern visitors; but seven hundred years ago this unusual structure—one of the few freestanding ruins still remaining in the Southwest—stood fully exposed to the burning sun. Yet the building was remarkably well adapted to its harsh environment. Its designers apparently understood the basic elements of both structural and solar design (see Figure 50).

Casa Grande was probably built around A.D. 1300 by the Hohokam, a sedentary agricultural people from whom the modern Indian tribes of southern Arizona, including the Papagos and Pimas, have descended. (To the Pimas, the name *Hohokam* means "all used up"; archaeologists prefer to translate the term as "those who have gone.")

Between 300 B.C. and A.D. 1450, the Hohokam controlled most of the northern Sonoran desert from Flagstaff south to Tucson in a loose confederation joined more by language and tradition than any political bonds. Originally nomadic hunters and gatherers who may have migrated north from Mexico, the Hohokam settled in this area during a period of relatively benign climate. Still, rainfall was erratic and water was precious, and the Hohokam tapped the meager flows of the Verde, Salt, Gila, and Santa Cruz rivers and diverted thin streams into their farmland. Indeed, sometime around A.D. 1100 the environment deteriorated so badly that the Hohokam economic and political systems were disrupted and trade with Mesoamerica essentially ended. In an apparent attempt to reorganize the population, the Hohokam leaders spurred the construction of ceremonial mounds, reminiscent of the Mexican structures. By 1300, the population level at least had stabilized enough for the Hohokam to build large structures such as Casa Grande.

Unlike their contemporaries, the Anasazi and the Mogollon, who built homes and entire villages into the natural shelters of canyon walls and high cliffs, the Hohokam generally remained on the ground in their flat desert and valley lands, creating structures especially adapted to that environment. At first these dwellings were simple pit houses, no more than holes scooped from the ground and covered over with low brush roofs. Gradually they de-

Figure 50. *The Casa Grande National Monument south of Phoenix, Arizona. The steel canopy protects the fragile adobe structure from infrequent rains.*

veloped more elaborate and, by contemporary standards, advanced dwellings: freestanding multistoried buildings of rock, clay, and adobe brick. Often several such buildings would be clustered together in a central complex enclosed by walls.

Casa Grande, or the "big house," was probably built as a dwelling for Hohokam leaders and served as the focal point for an extensive social and ceremonial center. Many smaller structures, now no more than low, rounded foundation walls, surrounded the big house; and archaeologists suspect that a host of other, less permanent dwellings spread out into the desert. The traces of an elaborate irrigation and drainage network can also be found nearby. Obviously, these ditches—and the construction techniques embodied in the big house itself—were direct responses to harsh climatic conditions.

The main building is a rectangle with its long axis running east–west. The walls are constructed from a mixture of adobe brick and caliche clay. (Found throughout this part of Arizona, caliche has the properties of natural cement: when pulverized and remixed with water, it dries into an extremely hard and durable surface.) Although the building is three stories high, the first floor is really a platformlike substructure. The floor plan is a symmetrical pattern of five rooms, with four rooms of equal size, each two stories high, surrounding a central room three stories high. The interior ceilings and the roof were constructed by laying long poles across the walls so that the ends protruded beyond the wall line. Several layers of overlapping branches, plus mats of woven reeds, were then laid over the poles and these too were covered with clay. Doorways were constructed as part of the design, with dressed logs serving as jambs and lintels. The windows and also several apertures on the west façade were apparently punched out of the solid wall after construction was completed. The designers also understood something about structural stress, for the high clay walls gently taper inward as they rise, giving them a slightly bowed configuration when seen in cross section and reducing the strain on the upper levels. The technique also creates the illusion (intentional, perhaps?) of an even taller structure.

Like all good designers, Casa Grande's builders incorporated solar potential into their structure. The thick walls and tiny windows provided good insulation in both summer and winter. The protruding logs of the roof could have been hung with blankets or mats to provide shade in summer, or removed in winter to allow maximum sunshine.

But the builders of Casa Grande may have had an even more intimate connection with the sun. The windows and the strange holes in the second- and third-story walls seem deliberately constructed to observe the sun. Obviously, the openings were carefully made; once poked out of the completed walls, their edges were smoothed and plastered over. And one such window was later completely filled in, as if some mistake in calculation had been made (see Figure 51).

John P. Molloy of the University of Arizona has found that eight of the fourteen openings at Casa Grande are aligned with risings and settings of the sun at both the solstices and the equinoxes. Molloy thinks the method of alignment was the "alternate door jamb" technique, similar to that used by the Mayas in the Caracol at Chichén-Itzá. An observer would sight along the line formed by the inside edge of one doorway or window and the outside edge of a second doorway or window. Such a sightline could be used to mark the passage of an astronomical object with some precision. In the case of a large object, such as the sun or moon, the moment when the limb or leading edge of the body first touched the line could be considered the first sighting. In addition, the sun might have shone through the round ports to strike a design or marker on an interior wall, thus signifying special days in the year.

At Casa Grande, as at the other Hohokam desert habitations, adaptation to the environment was vital for survival. The construction of buildings from adobe and caliche was one response; the development of water-control systems was another. However, even more basic was development of the means for predicting seasonal changes so seeds could be planted at the proper time, irrigation canals could be cleared and prepared for rains, and crops could be harvested before they were burned away in the sun. To do this, some sort of calendar was necessary. And like most other peoples, the Hohokam found their seasonal clock in the sky.

148

Figure 51. *The west façade of the Hohokam ruins of Casa Grande, showing a doorway and three "portholes" thought to have been used for astronomical observations.*

One of the more exciting developments in archaeology during the past two decades has been the realization that the American Indians had developed relatively advanced cultures. Scattered across the vast North American continent in hundreds of small tribal groups, speaking a host of different languages, with each a separate society shaped in large part by the local environment, the Indians who inhabited the area north of the Rio Grande were long overshadowed by the more sophisticated civilizations of the south.

With the exception of the Iroquois Confederacy in the northeastern woodlands, the Amerindians seemed disorganized, unimaginative, unintelligent, and existing only one step above Stone Age savages. Thanks in part to the techniques of the so-called new archaeology, which draws on contributions from a variety of scientific disciplines, a revised image of these first settlers in the New World is now emerging. Rather than being viewed as simply nomadic hunters, the Amerindians are now being reevaluated as budding technocrats who adapted to environmental conditions in quite resourceful ways, carried on extensive trade and intertribal relations over great distances, and developed considerable artistic and technical skills. One of the disciplines contributing much to this new appraisal of the Amerindian is astronomy, or more precisely, archaeoastronomy. Equally important, by discovering remarkable similarities in the cosmological viewpoints of these widely separated people, archaeoastronomy is providing further clues to the evolution of New World cultures and helping to map the spread of technological and scientific thought throughout all of the Americas.

The finding of a technical sophistication among the Amerindians parallels in part recent discoveries that have pushed back the estimated arrival time for these immigrants to the New World. Most evidence now supports the theory that the first humans crossed from Asia into the Americas as early as forty thousand and perhaps even as early as seventy thousand years ago. As recently as the 1930s, however, man's first presence in the Americas was thought to go back no more than ten thousand or twelve thousand years. This estimate was based on the discovery of the bones of a butchered mammoth found in 1932 near Clovis, New Mexico (the distinctive style and shape of the axe and knife blades found at this site were given the name Clovis points).

In 1958, however, charcoal found near Lewisville, Texas, at a site littered with stone axe heads was radiocarbon dated at thirty-seven thousand years. The results of this technique, which involves measuring the half-life of radioactive isotopes of carbon, are not always foolproof, so most archaeologists withheld judgment on the finding. Then in 1973 a piece of caribou bone, apparently carved and worked into a point as a crude tool, was found along the Old Crow River in the Yukon. Its radiocarbon date made it some twenty-nine thousand years old. The same year, a human skull found near Del Mar, California, was found to have a radiocarbon dating of more than forty thousand years.

Two years later, at Pittsburgh, Pennsylvania, the remains of a human habitation in a rock shelter showed a radiocarbon age of some nineteen thousand years. And in 1977 the remains of a butchered dwarf mammoth were found on Santa Rosa Island off the coast of California. The age of the charcoal chunks in the ancient campfire again exceeded the forty-thousand-year effective limit of the radiocarbon dating technique.

This accumulation of earlier and earlier dates, along with information on the blood types of current Indian groups in the Americas, suggests that successive waves of immigrants crossed the Bering Strait, either during glacial periods when the oceans receded enough to create landbridges or at other times, when people simply walked over the frozen waters. The first crossing was surely more than forty thousand years ago; it could have been even thirty thousand years earlier.

The growing evidence for an early arrival supports the premise that these immigrants to the Americas would have had ample time to develop independently the relatively advanced cultures found by the first Europeans. Moreover, the development of astronomical awareness and basic observational techniques now appears to have been a natural, almost necessary evolutionary step. Studies of the North American Indians—allowing for some cross-cultural interchange—offer a rare opportunity to measure the step-by-step development of astronomy (and science) among preliterate people apparently unaffected by outside influences. This evolution obviously reached its peak with the Mayas' intricate intertwining of astronomy, religion, and politics.

Research has concentrated on four major areas in North America: the solar alignments incorporated into the permanent structures built by the agricultural societies of the Southwest; the similar alignments found in the stone rings and wood rings of the Plains; the earthen mounds of the Mississippi and Ohio valleys; and the rock art found in the Southwest.

Most native North American astronomy was similar in its simplicity and function to that of other early preliterate societies, and was concerned primarily with the sun. The sunrises and sunsets of the equinoxes and solstices, as well as the sun's zenith passages, were used to mark dates of religious or agricultural importance. A lunar calendar was maintained by counting phases rather than by marking horizon positions. However, the planets, and the brighter stars such as Sirius and Aldebaran, were watched as they rose and set along the horizon, usually in conjunction with the sun and moon, to mark divisions of the year.

"The use of the horizon observations as means of regulating the calendar is best adapted to a sedentary society that has an opportunity to identify suitable local landmarks denoting the seasons or to erect markers," writes Stephen C. McClusky in his article "The Astronomy of the Hopi Indians." "Nomadic or cosmopolitan societies tend to use an astronomical system that is valid over an extended geographic area, hence the concern of the Greeks and Egyptians with the heliacal risings of stars as a means of regulating the calendar. Since the calendar of an expanding society is almost certain to be portable, it is not likely to be based upon horizon observations."

By a lucky quirk of history, many Indian societies of the American Southwest were not only sedentary, they were also nearly permanent over a long period extending from before the time of Christ until the present day.

Moreover, perhaps because of the relatively inhospitable nature of the land, the Hopis and other Southwestern Pueblo tribes remained virtually untouched by European influences until late in the nineteenth century. Thus it is possible both to find traces of astronomical systems incorporated into ancient ruins and to find applications of the old observing techniques and calendric devices in contemporary folklore and religion. While most Southwestern tribes now follow the national civil calendar, the old festivals and feasts of the past are still observed. Indeed, many of the Roman Catholic "saint's days" celebrated by the eastern Pueblo peoples of the Rio Grande Valley are related to older Indian holidays still observed by the Hopis. These festivals, mainly solar celebrations, including the Wuwuchim, Soyal, and Niman Kachina, required some observational precision in sun-watching to ensure they occurred on the proper date each year.

To establish these dates, the Hopis used a luni-solar calendar based on two rather simple observations: They watched for the northernmost and southernmost settings of the sun on the western horizon to determine the solstices, and for the first appearance of the thin crescent moon at sunset following the solstice to signal the beginning of a month count. "These two simple observations alone can be used to regulate a luni-solar calendar, although knowledge of further regularities of the motion of the sun along the western horizon and of the interval between successive new moons could improve reliability," writes McClusky. "But what is important is that we need nothing more than a simple set of observations in order to establish the beginning of a specific month. There is no need for any theoretical formulation of the relationship between the sun and the moon."

The system was not quite that simple, however, because the interval between the solstice and the appearance of the crescent moon constantly changes. In most years, when the solstice occurs in conjunction with a twelve-month lunar cycle, this interval decreases by eleven days, so that ceremonial dates come eleven days earlier each month. In those years when the solstice comes after thirteen lunar months, the interval is closer to nineteen days. In the end, the Hopis developed what McClusky describes as a "framework in which the heavenly bodies are seen as running alternately fast or slow—a concept diametrically opposed to the concept of uniform motion in the heavens that played such a major role in the development of Western astronomical thought."

To accommodate the reality of their observations, the Hopis apparently developed a system for charting the cycles of lunar and solar positions and phases. This was intended to explain the changing intervals and to mark the times of their ceremonies with considerable precision. As a result, the dates of festivals were never more than a day or two off in the Hopi calendar.

Interestingly, the astronomical knowledge allowing this precision was not restricted to a single priestly caste as in most preliterate societies. Although the "sun-watchers" formed a special class within the tribe, the essential observational information was widely disseminated among other people of the pueblo, and a negligent sun-watcher could be severely criticized for his poor work. In fact, the Hopis apparently determined the exact date of a festival as much by consensus as by decree; alternative observations by other, nonofficial sun-watchers could sometimes overrule decisions by the official observers. Moreover, the elaborate sequence of festivals, each with its own observational markers, served as periodic double-checks on the total calendar and offered a backup for any observations that may have failed because of bad weather or human sighting errors.

The Hopi calendar was completely integrated with the concepts of fertility, birth, and growth. It was essential that rituals be performed faithfully and that observations of the sun be made with extreme care; otherwise the crops would fail. "For a society like the Hopi, subsisting in a marginal agricultural region, the maintenance of their calendar is absolutely essential to survival," says McClusky. "The complexity and redundancy of their astronomical system can thus be seen as an integral part of their social adaptation to this harsh environment."

The survival of the ancient Hopi calendar and the traditions of sun-watching have provided archaeoastronomers with the clues for finding traces of earlier observational systems. Most often these clues are preserved in the crumbling remains of once-thriving settlements scattered throughout the Southwest. The sun-watching sites are usually of two types, as classified by Ray Williamson of the University of Maryland. Type 1 consists of man-made structures generally used in conjunction with some natural feature on the horizon to gain information about the sun's motion for determining a calendar. Type 2 consists of man-made structures designed so that certain astronomical phenomena will become obvious through the interplay of light and shadow—for example, sunlight passing through an aperture on one wall to illuminate a marker on another.

An example of the type 1 sun-watching site can be found at the Wijii ruins in Chaco Canyon, New Mexico, where there is a white sun symbol painted at the top of a stone stairway. Some five hundred meters to the east-southeast, in the direction of the midwinter sunrise, there is a natural cleft in the stone face of a high butte rising over the horizon. Because the cleft is narrower than the diameter of the sun, an observer standing at the "sun painting" and watching the sun rise in the gap would be able to determine the date of the solstice within a day or two.

154

Casa Grande is an example of type 2, where windows and ports were deliberately incorporated into the building for specific astronomical observations. A similar architectural device can be seen in the "castle" building at the Hovenweep ruins of Utah. Here, the so-called D tower was built around A.D. 1200, with other rooms added later. Among the additions is a small chamber with a doorway through which the setting sun at the summer solstice shines briefly on the interior wall. Although the ruins leave but scant evidence, the builders may have also created special ports, or sighting holes, in the walls, one aligned to the summer solstice and the other to the winter solstice. In addition, the outside doors were aligned so their jambs (using the same alternate-jamb observation method as the Mayas) pointed to the sunsets on the two equinoxes.

The Hovenweep ruins are remains of a people called the Anasazi, contemporaries of the Hohokam, who inhabited the northern Colorado plateau, or roughly the Four Corners region where Arizona, New Mexico, Utah, and Colorado share a common boundary. The Anasazi, or the "Ancient Ones," were also struggling agriculturalists; and before the climate changes in the late 1300s they built elaborate cliff villages and stone pueblos throughout northeast Arizona and northwest New Mexico. Their descendants are the Hopis and the other Pueblo tribes of the upper Rio Grande Valley.

The largest Anasazi settlement—indeed, the largest settlement in all the Southwest during this epoch—was Pueblo Bonito, now a well-tended ruin in the Chaco Canyon National Monument of New Mexico. Pueblo Bonito, which once housed as many as six thousand people, reached its peak between A.D. 950 and 1150, when it probably served as a trade distribution center for the other settlements spread throughout the area. In keeping with its importance for Anasazi culture, Pueblo Bonito and the other ruins of Chaco Canyon have provided a wealth of clues to Anasazi astronomy.

Pueblo Bonito has a strange unsymmetrical shape, with its ground plan looking almost like a giant letter D laid on an east–west axis so the straight back of the D forms a high wall along the north side (see Figure 52). Ray Williamson suggests the shape and

155

layout were deliberately conceived to make maximum use of solar energy. The great semicircular shape open to the south is an effective solar collector, with the high northern walls both reflecting the sun's rays into the central plaza in winter and protecting it from the prevailing northwest winds. Moreover, the interior rooms act as efficient moderators of the wide temperature extremes experienced in Chaco Canyon.

Jonathan Reyman of Illinois State thinks there may have been alignments between the exterior corner windows of Pueblo Bonito

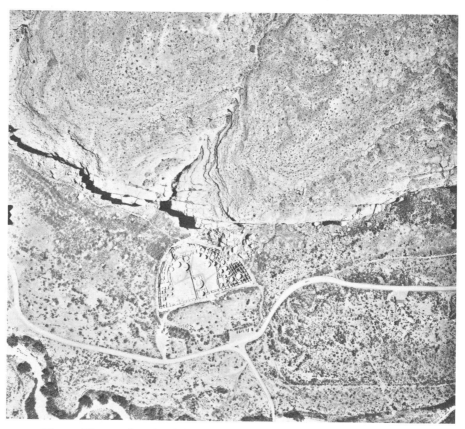

Figure 52. *Aerial view of the ancient Anasazi settlement known as Pueblo Bonito. Note the distinctive* D *shape of the building complex.* Photo courtesy U.S. National Park Service

and the winter solstice sunrise. As he notes in an article for *Science,* "a record of this event was necessary for the establishment of a solar calendar, and possession of an accurate calendar increased the Chacoans' chances for a successful harvest."

Casa Rinconada, one of the great kivas (ceremonial rooms) at Pueblo Bonito, has a wall niche that is illuminated by the sun shining through a northeast window during a period of four or five days around the summer solstice. Originally this circular kiva was covered by a massive roof supported by four posts at points along the rim corresponding to the semicardinal directions (northeast, southeast, southwest, and northwest), apparently marking the yearly excursions of the sunrise and sunset points. (The modern Hopis, incidentally, still consider these points as important and see them as the lines intersecting at the center of the universe.) In a roofed-over and darkened kiva, when the midsummer sun shone through the northeast window to illuminate the niche on the opposite wall, the effect must have been striking. (Williamson, who has studied the solstice alignment at Casa Rinconada, suggests that this light show may have been related to a Pueblo myth about a maiden impregnated by the sun while she lay asleep in front of a small window.) Further, this kiva was placed due west of a major shrine located on top of a mesa. At dawn on the equinoxes, the mesa's edge cast a shadow cutting the kiva in half, with the band of darkness slicing diagonally through its exact center.

Chaco Canyon is also the site of what has been called "the most exciting discovery in North American archaeoastronomy," a discovery that suggests an even greater sophistication among the ancient sun-watchers. Generally, most North American peoples followed horizon-based astronomy systems—that is, they were concerned with the rising and setting of objects against natural or man-made markers on the horizon. This was in contrast to the zenith-oriented astronomy of the peoples in South and Central America, the pole-oriented astronomy of the Chinese, and the ecliptic- or zodiac-oriented astronomy of the Babylonians. Yet one distinctive marker found at Chaco Canyon indicates that the Anasazi may have also observed and held important the zenith passage of the noon sun on the day of the summer solstice. Moreover, as Kendrick Frazier

Figure 53. *Diagram of the changing light and shadow pattern at the solar construct on Fajada Butte.* Left: *On the summer solstice, a dagger of light moves through the center of the larger spiral;* middle: *On the autumn and spring equinoxes, one streak of light bisects the smaller spiral;* right: *On the winter solstice, the two streaks of light "frame" the larger spiral.* Illustrations by the author, after Sofaer, Zinser, and Sinclair

wrote in *Science News,* the configuration of this particular alignment and marker "may possibly indicate a sophisticated understanding by the ancient peoples of Chaco Canyon of the interplay of light and angled surfaces to achieve an observed event."

On June 29, 1977, Anna Sofaer, a Washington-based artist with an interest in prehistoric Indian art, climbed the 430-foot Fajada Butte in Chaco Canyon to photograph two interesting—but apparently meaningless—petroglyphs of spiral design carved high on the butte's east face. To reach the carvings meant hiking up from the canyon floor, treading carefully past rattlesnakes sunning along the upper ledges, and descending a thirty-foot natural chimney. Finally she had to squeeze behind three massive slabs of rock— each approximately nine feet high, six feet wide, and one foot thick—that leaned upright against the butte wall. Scratched into the wall just behind the rocks were two slightly elliptical spiral designs. One was about fourteen inches in diameter with nine rings, or turns, the other about five inches wide with two and a half turns, plus a loop extending to its right.

As Sofaer knelt to photograph the petroglyphs—at noon, as it so happened—she noticed a bright sliver of light in the shape of a thin dagger suddenly begin moving vertically down the larger spiral just to the right of its center. The light beam, which moved completely through the spiral before disappearing, was created by the sun's rays penetrating a small gap between two of the three stones leaning overhead (see Figure 53).

158

"It was obvious that the light meant something," Sofaer later told an audience at the Los Alamos Laboratory in New Mexico. "Right away I thought of the solstice. It was a week after the solstice and it occurred to me that the spiral was put there to record it. A week earlier the light would have passed through the center. It was an incredible coincidence that I was there just a few days past the summer solstice, at noon. If I had come a little later or earlier, I would have missed the whole thing."

Sofaer returned to Fajada Butte the following year accompanied by Rolf M. Sinclair, a physicist with the National Science Foundation, and Volker Zinser, an architect who had studied the effects of light and shadow caused by different types of curved and angular shapes. In mid-May they photographed the light patterns playing over the spirals at local solar noon on successive days and identified the particular surfaces of the rock slabs that were responsible for creating the dagger of light. Preliminary measurements indicated that on the day of the solstice, the sun, as it moved west, would cause the dagger of light to move down the rock wall and through the spiral's center. And so on June 21, 1978, they set up cameras and carefully recorded the downward movement of the light. It passed directly through the center of the spiral as predicted. Moreover, the light pattern remained consistently dagger-shaped and maintained its motion for eighteen minutes.

The observers also found that a second, much smaller, and almost imperceptible spot of light formed by another gap in the overhead rock slabs appeared to the left of the larger spiral on the solstice. At the autumn equinox, September 21, this second light pattern changed from a spot to an elongated triangle and moved directly through the center of the smaller spiral. At the same time, the larger dagger moved to the right of the larger spiral. By the time of the winter solstice, December 21, both bands of light shifted to the right; and at noon they moved downward on either side of the large spiral to "frame" it perfectly. At the spring equinox, March 21, the pattern shown in the fall was repeated.

A device to mark the divisions of the solar year by means of the midday position of the sun is so far unique in American Indian astronomy. All other devices use the rising and setting points of the sun on the horizon. However, it is still unclear how much of the complex light show on top of Fajada Butte is intentional. "The natural appearance of this solar construct and its integration with the surrounding environment makes man's role perceptible only through careful examination of all the dynamics operating to form the significant solar markings," wrote Sofaer and Zinser in *Science* magazine.

For example, the three rock slabs possibly could have fallen from an overhang some twelve feet above the ledge. Then some Anasazi who discovered the light pattern could have carved a spiral petroglyph to take advantage of the natural solstice marker. But geologists who have inspected the site doubt that the rock slabs fell exactly into place because the original position of the stones was too low and, more important, off to the side of the present location. Sofaer herself thinks the rocks were deliberately placed in position to capture the light, and worked carefully to create the precise dagger-shaped beam that bisects the spiral.

According to Zinser's studies, the curves of the surfaces on the leaning rocks are crucial. They need precisely the right shape and angle to maintain the thin dagger shape as it moves vertically down the spiral. Zinser thinks the rock slabs were worked with tools to produce the proper angles, but his premise is not easily tested since erosion of the soft sandstone has worn away any definite signs of tool marks.

Of course there is no way to prove exactly how the light show was created, yet it seems highly unlikely it was pure coincidence and just a quirk of nature. Other evidence of solstice-watching among the Anasazi certainly supports the theory that they had the intelligence, desire, and ability to produce such devices, particularly those incorporating natural elements. "The monumental quality of this solar construct reflects the profound beauty of ancient Pueblo architecture," writes Kendrick Frazier in *Science News*. "It is characterized by the Indians' sensitive integration of their structures with nature, light, and patterns of the solar cycle."

Despite their close harmony with nature, it may have been nature itself that destroyed the Anasazi and Hohokam cultures. A few hundred years before the arrival of the Spanish in the New World, the great societies of the Southwest disintegrated and dispersed.

By the time Coronado and other explorers trekked into the Rio Grande Valley, these peoples had disappeared completely, leaving few clues to explain the collapse of their civilization. Some specialists suspect a widespread climatic change so reduced the food production that the social units broke up and the great settlements simply crumbled into dust as the people either died from disease and starvation or moved into smaller pueblos to the south and west. Still, the downfall remains a mystery.

Another long-term mystery in archaeology has been the origin and function of the massive earthworks found throughout the Mississippi River basin. Between 1000 B.C. and A.D. 1300, a loose cultural group controlled an area that stretched from the Great Lakes to the Gulf of Mexico, leaving an enigmatic record of their existence in dirt mounds.

Interest in—and understanding of—these peoples, generally referred to as the Mound Builders, has been renewed recently because of research conducted by archaeoastronomers. Their discoveries are leading to a new appreciation of this culture and providing some clues to the possible transmission of information and technology between the Mississippi Valley and the Valley of Mexico.

The early colonists who crossed the Appalachian Mountains into the Mississippi basin found thousands of strange earth mounds in the valleys and plains created by the river system. Some were cone-shaped and as large as small hills, others were squat and flat-topped. Some were laid out as giant abstract designs: zigzag lines and spirals sprawling over several hundred yards. Still others were huge earthen effigies of animals or humans. The mounds were a source of wonderment for the European settlers. The possibility that they had been built by Indians was immediately dismissed, for the Indian was considered too savage, too backward to have created anything so grand. As Brian Fagan has noted in his *Elusive Treasure*, "the very primitiveness of the American Indians compounded a puzzlement over their origins. The North American landscape was devoid of any conspicuous monuments of antiquity, of any lasting memorials of ancient civilizations that had preceded the Indian or lived alongside of them."

The discovery of the mounds, therefore, served as "evidence" of that earlier, higher culture the white settlers so desperately sought. They were convinced the mounds were the work of an advanced civilization, or people of foreign origin who had once occupied this land before being displaced by the savages. Thus the mounds were credited variously to the Egyptians, the Phoenicians, and even the Lost Tribes of Israel. One school ascribed the earthworks to survivors of the sunken continent of Atlantis. And among Americans of Celtic descent the mounds were attributed to Saint Brendan or the legendary King Madoc and his wandering band of Welshmen. (Ironically, there is some slight similarity between certain Indian mounds and the long barrows, earth rings, courses, and human effigies found throughout the British Isles.) In short, as Fagan notes, "the Mound Builders appealed in the nineteenth century to the same interests and emotions that the Ancient Astronauts and extraterrestrial visitors arouse today."

Even after American archaeologists recognized and described the true origin of these mounds, their significance was not fully appreciated. Amateurish attempts to recover "buried treasure" led to the wholesale destruction of many of them, usually with complete disregard for dating and cataloging the artifacts found within them. Moreover, since the mounds were often considered only more primitive versions of the structures in Mesoamerica, they were dismissed as unworthy of either scholarly investigation or governmental protection. Only in the last half-century—and really the past decade—have the mounds been revealed as the visible remains of a dynamic and sophisticated culture.

The mounds actually were built by several different groups that succeeded each other over the period from nearly 1000 B.C. to A.D. 1400. The first of these societies was the Adena, an agricultural people living in small villages scattered along the Mississippi river system from Canada to the Gulf of Mexico. By 300 or 200 B.C., as part of an elaborate burial ritual for their leaders, the Adena began constructing mound complexes.

Extensive digs at Koster and other sites in the lower Illinois Valley have shown that the Adena people and their predecessors, somewhat contrary to the popular image of restless nomads, maintained long-term habitations. Layer upon layer of artifacts at these sites demonstrates both a sense of permanence and a thoughtful adaptation to the particular environment.

Around A.D. 200 the Adena society was replaced by a new socioreligious group, the Hopewell people. The Hopewell earthworks were even more spectacular and grandiose, with structures in the shapes of circles, octagons, and serpentine lines making up huge, sprawling ceremonial centers. Although remaining centered in the Ohio Valley, the Hopewell people controlled or influenced (mainly through trade) a vast portion of North America. Their main trade routes ran east to the woodlands of the Iroquois, but contacts also extended north, west, and south. The common currency of the culture was obsidian, a hard glasslike black rock quarried from outcroppings near today's Yellowstone National Park, more than a thousand miles west of the Ohio Valley. Shells from the Gulf of Mexico were used as necklaces and bracelets. Mica from the southern Appalachian Mountains was cut into stunning silhouette shapes of hands, birds, and human figures. Copper nuggets from the western Great Lakes were beaten into breastplates, ear plugs, and knife blades.

Sometime around A.D. 800–900, a new influence—possibly from Mexico—spread into the Mississippi Valley. Another culture, known simply as the Mississippian, began building large temple mounds, usually grouped around a central plaza in the style of Teotihuacán. Serving the living rather than the dead, these earthworks were used both as dwellings for chieftains and as ceremonial centers for political and religious activities. The largest and grandest of the known Mississippian cities was at Cahokia, located across the river from present-day St. Louis in southern Illinois. In A.D. 1200, at the peak of the late Mississippian culture, Cahokia may have housed as many as ten thousand to thirty thousand people. More than a hundred mounds were located within this urban complex and trade center. The largest, the so-called Monks Mound, towered one hundred feet in the air and covered over fourteen acres on the ground, making it nearly as large as any of the Mexican pyramids. Ann Daniel-Hartung suspects that this mound and the others were built almost exclusively by women workers, each carrying forty-pound earth-filled baskets.

Alas, many sites such as Cahokia have simply disappeared under the relentless spread of modern urbanization. Once-elaborate ancient towns are now buried beneath parking lots, apartment houses, shopping centers, highway interchanges, and even, in one case, a municipal golf course. One mound complex from the earlier Hopewell Period still exists at Newark, Ohio, however. It features a large square enclosure encompassing an area of some twenty-seven acres and an adjacent circular enclosure of similar dimensions joined to it by an avenue. Both of the enclosures were surrounded by various other earthworks, ranging from five to thirty feet high. The Newark site also has an octagonal structure, 625 feet to a side. At the junction of each side there are seventeen-foot-wide openings. Various square enclosures at Newark also have the same corner openings.

Archaeologists have no real clues to the possible uses of these structures. But obviously their construction required considerable surveying skill, knowledge of basic geometry, and a good deal of community motivation and organization. John Eddy has suggested they may represent ceremonial ball courts or playing fields, somehow related to the ball courts found in Mesoamerica and perhaps used for a form of lacrosse.

A further link to Mexico is suggested by James A. Marshall, a civil engineer who has carefully surveyed the mounds and concluded that the Hopewell people used a 187-foot unit of measurement in common with the builders of Teotihuacán. Marshall found that one side of the great square at Newark measured exactly five of these units; while the "liberty square" (another structure, in Ross County, Ohio) measured six units. He also claims to have found a number of smaller residential structures measuring one unit per side. Intriguing as his proposal sounds, most archaeologists are a bit leery of any "universal measurements," for the "Teotihuacán unit" sounds as questionable as the "pyramid inch."

A more promising lead to the mathematical, technical, and astronomical skills of the Mound Builders has been found at Cahokia. In addition to the mounds, the site still shows the remains of several circles formed by stockades, or rings of posts. The largest

164

Figure 54. *Aerial view of the Octagon Earthworks at Newark, Ohio, maintained as a state memorial by the Ohio Historical Society. An octagonal mound system encloses fifty acres and is joined by parallel mounds to a circular structure enclosing an additional twenty acres. The golf course of the Moundbuilders Country Club is laid out over and around the Octagon Mounds.* Photo courtesy Newark Area Chamber of Commerce, Newark, Ohio

of these "posthole circles," a ring 410 feet in diameter, has been studied by Warren L. Wittry, who dubbed it an American Woodhenge.

Originally, Woodhenge may have consisted of some 48 posts of indefinite height, each separated from the next by 7.5 degrees. A forty-ninth post was located at the center of the circle, offset about five feet from the true center. In 1977, Wittry, assuming from the hole size that the original posts might have the approximate diameter and height of modern telephone poles, placed three such posts on the circle's eastern rim. Then he raised another utility pole at

165

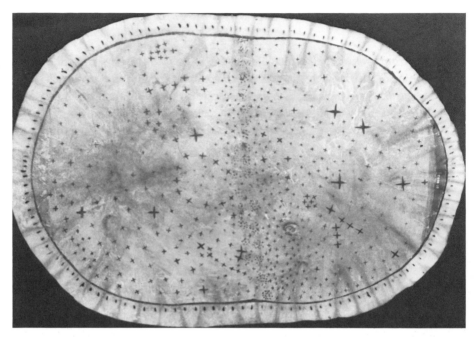

Figure 55. *The Pawnee buckskin star chart in the collection of the Field Museum of Natural History, Chicago. The chart is a single piece of thin, soft leather* (56 × 38cm) *painted with at least three pigments. A lacing (now missing) that once ran around the border apparently was used to draw the chart into a bag, perhaps for carrying a meteorite or other sacred items.* Photo courtesy Von Del Chamberlain

rings, unrelated to encampments. These are large circular structures, sometimes with huge central cairns and lines, or "arms," of stones laid on the ground and radiating out like the spokes of a wheel. The early white pioneers in the west, struck perhaps by the odd shape and imposing size of these rings, called them medicine wheels, implying some magical or mystical use. Recently, thanks primarily to the work of John Eddy of the High Altitude Observatory, some of the medicine wheels have been identified as possible astronomical observing devices.

Eddy began his search for astronomical alignments in the medicine wheels after visiting one of the better-known examples, the Big Horn Medicine Wheel in Wyoming. Located at an altitude of 9,640 feet on the flat shoulder of Medicine Mountain in the Bighorn range, this wheel is a crude circle of stones some eighty feet in diameter. There is a central cairn—actually, a concave pile of rocks—several feet high and about twelve feet in diameter. A series of twenty-eight spokes extends from the central cairn to an outer circle of stones. Along the outer rim there are six other U-shaped cairns of various sizes, five outside and one within. The dendrochronology dating of wood found at the site suggests it was built around A.D. 1760; and its existence has been known—to modern Americans, at least—since about 1880, when it was first discovered by prospectors. Today, Big Horn Medicine Wheel is under the trusteeship of the National Forest Service (see Figures 56 and 57).

During field research at the site in 1972 and 1973, Eddy found that by standing at the distinctive cairn that lies outside the ring and sighting back across the wheel and through the center of the hublike cairn, he was looking at the point on the horizon where the sun rises on the day of the summer solstice. By standing at a second cairn, and again sighting across the center, he could see the point where the sun sets on the same day. He suggests the builders did this intentionally, so they would have a second chance to mark the day if dawn proved cloudy.

Eddy found no alignments with other cairns or spokes that could mark the winter solstice. However, considering the inaccessibility of this particular site in the winter months, this is not surprising. What was surprising was his discovery that three other alignments marked the heliacal rising of three stars: Aldebaran, Rigel, and Sirius. Since all three stars are located in the same area of the sky as the sun at midsummer, they could have been used as precursors, or advance warning signals, of the approaching solstice or other specific dates. In fact, the heliacal rising of Aldebaran corresponded almost exactly with the summer solstice sunrise at the epoch when the medicine wheel was built.

Writing in *Technology Review*, this is how Eddy described what the dawn watchers on Medicine Mountain might have seen:

> About an hour before dawn, Aldebaran would rise. The predawn sky would already be blue, and all the dim stars would be gone. Indeed, the coming sun would be brightening the sky so rapidly that on this particular day Aldebaran would

Figure 56. *Aerial view of the Big Horn Medicine Wheel in Wyoming.* Photo courtesy USDA Forest Service

Figure 57. *A ground-level photograph of the Big Horn Medicine Wheel, taken by the Forest Service in 1935, when it was described as "an unsolved mystery." A stone fence that surrounded the site then has since been removed.* Photo courtesy USDA Forest Service

flash out like a beacon near the horizon, lasting only a matter of minutes before disappearing in the predawn glare. That phenomenon would make this day a distinctive one, for on the previous day Aldebaran would not have been seen at all [the sun's light would have masked it] and on the day after it would have persisted far longer. In short, watching for Aldebaran's flash at dawn would have given a precise indication of the solstice, accurate to within a day or two.

Twenty-eight days after the heliacal rising of Aldebaran, Rigel would repeat the phenomenon as seen from another alignment. And twenty-eight days after that, Sirius would experience heliacal rising. Eddy cannot draw any direct relationship between the twenty-eight-day cycle and the ring's twenty-eight stone spokes; it is, however, an interesting coincidence. He does suggest that the rising of Sirius could have been a signal to leave this cold and windswept mountain, for winter would be approaching soon thereafter. Indeed, this site can only be reached easily in the summertime; snow covers it much of the rest of the year, beginning as early as late August and continuing sometimes until mid-June. Although the site was practical for summer solstice observations, according to Eddy "the choice of a cold, arduously reached mountain top in preference to the equally usable nearby plains must be justified on other grounds—purely mystical or purely aesthetic."

To check his theory that the Big Horn Medicine Wheel was deliberately aligned with astronomical objects, Eddy made aerial and ground surveys of twenty other sites in the western United States and Canada. (In fact, most of the known wheels lie in the Canadian prairie provinces of Alberta and Saskatchewan.) He found most rings quite dissimilar. They had apparently been built at different times by different peoples and perhaps for different purposes. Yet by limiting his investigations to those rings that had distinctive spoke structures and not solely cairns, Eddy almost always found some alignment with astronomical objects. The most interesting site, and the one that had the most direct connection with the Big Horn wheel, was a medicine wheel on Moose Mountain (actually more a rolling hill) southeast of Regina, Saskatchewan.

This site has a large central cairn containing an estimated eighty tons of rocks, all of local origin. Five long spokes radiate out from the center and each ends in a smaller cairn. Although their configurations appear at first glance quite different, the Moose Mountain and Big Horn wheels share the same orientation, with the five cairns (plus a sixth small pile of stones unconnected to the central cairn) of Moose Mountain corresponding almost exactly to the layout of the Big Horn plan. As might be expected, then, the alignment of the outer cairn and hub marking the summer solstice sunrise was the same at both sites. In addition, alignments with the three stars seen over Medicine Mountain in Wyoming were duplicated at Moose Mountain.

However, the Canadian ring posed a problem. The alignments at Moose Mountain would work only if the site had been built much earlier, since precession slowly changes star positions. In fact, alignments seen in Wyoming around A.D. 1700 could only have been seen at the site in Saskatchewan around A.D. 100. (The sun's shift was minimal in this period, so the solstice alignments were not greatly affected.) Subsequent radiocarbon dating of charcoal fragments found at the Moose Mountain site confirmed that the wheel was indeed built—and used—some two thousand years ago!

The connection in time and space between the two medicine wheels provides still another example of the long tradition of practical astronomy among the North American Indians, including those supposedly restless and rootless wanderers of the Plains. As Eddy notes sadly, that tradition was apparently lost completely after the arrival of the first Europeans and the introduction of the horse sometime in the mid-1500s.

No oral history or lore supports the astronomical use of the medicine wheels and no records of their observations have been maintained. The possible function of the stone rings has been deduced solely by the analysis of their celestial alignments. We can only assume these rings were once the tools of astronomers.

By contrast, one set of Indian astronomical records does exist: rock art of the Southwest showing a possible supernova explosion. Archaeoastronomers only wish they knew how—and when—these observations were made.

Stars in Stone

"On the day kuei-hai *in the tenth month of the second year of the Chung-P'ing reign period, a guest star appeared within* Nanmên. *It was as big as half a mat; it was multicolored and it fluctuated. It gradually became smaller and disappeared in the sixth month of the year following the next year. According to the standard prognostication this means insurrection."*

—From a Chinese record, A.D. 185

Bathed in an eerie green glow from the television monitor, the faces of the scientists and technicians in the dimly lit room showed the tension and anticipation of all discoverers about to enter a new world. Line by line, the video image slowly filled the screen to reveal a strange, fanlike shape with a bright, distinct point at its center. The group gave a collective gasp and, then, after a burst of mutual congratulations, began an excited discussion of the various features revealed on the screen. The scientists, members of the imaging team on the "Einstein satellite," had just seen the first focused picture of the Crab nebula in X rays (see Figure 58).

Launched by the National Aeronautics and Space Administration in November 1978, the Einstein satellite (or, more officially, the HEAO-2, for the second in a series of High Energy Astronomy Observatories) carried the first telescope capable of producing actual pictures of celestial objects emitting X rays. Previous detectors

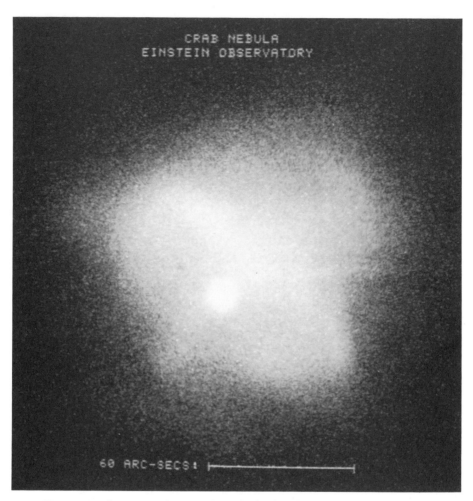

Figure 58. *This X-ray photograph of the Crab nebula taken by the Einstein (HEAO-2) satellite in 1979 shows both the expanding shell of gas and dust and the pulsar at its center, appearing here as a bright circular point.* Photo courtesy Harvard-Smithsonian Center for Astrophysics

had been capable of determining only the position and intensity of these objects. Within the initial three months of operation, the telescope, designed by scientists at the Harvard-Smithsonian Center for Astrophysics and operated by a consortium of four research institutions, had produced images of the brightest and most distant objects known to man—quasars some ten billion light-years

174

from earth. The telescope also produced X-ray images of galaxies, clusters of galaxies, globular clusters, and, most interesting to some researchers, the remnants of stars that had exploded eons earlier. Among these dead stars, now seen as vast expanding clouds of dust and supercharged gas, was the Crab.

The Crab nebula is one of the most fascinating objects in the universe, and, for modern astrophysics, it has become a touchstone: a cosmic reference point for studies concerning everything from the production of extremely high-energy radiation to the evolutionary processes that determine the birth and death of all stars.

The Crab nebula was first observed through optical telescopes over two hundred years ago. In 1758 the French astronomer Charles Messier, a diligent searcher for comets, discovered a fuzzy patch of light on the edge of the constellation Taurus. Although he considered this "smudge" something of a nuisance, he dutifully recorded the observation and described the object as a "nebulosity." He also designated it "Messier 1" in the catalog he began compiling, so such objects wouldn't again be confused with comets.

About a century later, Lord Rosse, the eccentric Anglo-Irish astronomer who built the then world's largest telescope on his estate in central Ireland, observed the same object and made a diagrammatic sketch. Through the telescope, the indistinct smudge resolved into a ragged shape with apparent appendages, or limbs, extending from a brighter central core. Perhaps seeing some resemblance to a crustacean, Lord Rosse dubbed it the Crab nebula.

The delicate, wispy structure of the Crab as revealed by modern time-exposure photographs is intrinsically beautiful and appealing (see Figure 59), but this elegant shape also holds the keys to understanding the underlying principles of modern astrophysics. Even the first identification of the Crab by Messier as something "different" from the familiar stars, planets, and comets indicated that other classes of celestial objects could exist in the universe. More important, the Crab's nature suggested that celestial objects might change—indeed, might even be destroyed—thus undermining the philosophical notion that the heavens were immutable and eternal.

Figure 59. *Optical view of the Crab nebula. The arrow points to the pulsar at the center of the nebula thought to be the remains of the original star.* Photo courtesy Smithsonian Astrophysical Observatory

When radio telescopes went into general use after World War II, one of the first objects seen as an emitter of radio noise was the Crab. Then in 1967 British astronomers running a routine test of their radio equipment picked up an odd frequency—a weak signal that rhythmically pulsated several times a second. The astronomers believed at first they either had mechanical difficulties or had somehow picked up earthly interference. After an extensive check, however, they came to the unmistakable conclusion that the rapidly fluctuating signal came from space: The first pulsar had been discovered.

Subsequently, more than 150 of these objects would be detected, some pulsing as rapidly as 30 or 40 times per second, each with a precise, dependable, natural radio beacon in space. Astronomers now believe a pulsar is the dense, compressed core of a collapsed star whose outer shell of gaseous material has been blown away by a supernova explosion. This tiny, extremely compact sphere of heavy elements is spinning rapidly and spewing out bursts of high-energy radiation with each revolution. Because the magnetic field around such a dense object is so strong, the radiation can escape only along the field lines—that is, in one direction. Thus, the flow of energy is seen as a narrow beam that sweeps past an observer like the beam of some cosmic lighthouse. The energy is so intense—and the revolution of the star so rapid—this radiation appears as quick repeated bursts or pulses.

Because the Crab nebula also resulted from a supernova explosion, astronomers suspected that the original star might still be present. Sure enough, when radio antennas were trained on the Crab, they revealed a systematic pulsing associated with a faint star buried deep within the center of the Crab's expanding cloud of dust and gas. Other observations in gamma and cosmic rays have shown the Crab's central star is a source of this high-energy radiation as well.

Finally, in early 1979 the Einstein satellite produced the first picture of the Crab nebula in X rays, a form of radiation not seen on earth because it is absorbed by the atmosphere. The X-ray pictures received with such excitement over the television monitors at the Center for Astrophysics showed a structure much different from the traditional optical view. The wispy filaments of material in the dusty gas shell had disappeared. Instead, the Crab showed an odd symmetry, with two fanlike rays spreading out from a central point, much like the bow front of an expanding wave. And, as suspected, a star—much brighter in X rays than in optical light—appeared at the center of the nebula. In timed-photo sequences, this star seemed to pulsate with an X-ray frequency comparable to the radio signal. There could be little doubt that this small central point was the original star left behind by the supernova explosion.

When the X-ray pictures of the Crab are placed side by side with radio and optical images, a remarkably complete anatomy of this unusual object is possible. The outward expansion of gas and dust can be measured and its velocity calculated. From the current size of this gaseous shell, astronomers now know it must have begun with an explosion some nine hundred years ago. More remarkable, such an explosion, appearing as a bright flash in the constellation Taurus, was actually seen and recorded by both Chinese and Japanese astronomers on the fourth of July in the year 1054. And new evidence, albeit largely circumstantial so far, suggests that American Indian sun-watchers scattered throughout the Southwest may also have seen the exploding star and recorded the observation in their rock art.

In short, just as the evolutionary history of stars may be expressed in the physical processes exhibited by the Crab nebula, so may the evolution of modern astronomy be recorded in humankind's observations of this celestial object.

Most stars in a galaxy like the Milky Way are remarkably stable. For millions of years, they will emit a steady outpouring of radiation with only slight variations. (Actually, our sun and others like it may be quite variable. But these variations are subtle and relatively short-term. While they may affect the quality of life on a planet in their immediate neighborhood, the variability may never be noticed on more cosmic time and distance scales.) Occasionally, however, perhaps once every hundred years in a given galaxy, a star reaches such an advanced state of old age that it burns off most of its nuclear fuel, and the delicate equilibrium that holds it together can no longer be maintained. Internal pressure overcomes the force of gravity and the star explodes in a brilliant and violent flash of light and gas. For a brief time, usually a week or a few months at the most, the star becomes a million times brighter and emits as much energy as all the other stars in its galaxy combined. The star's outer layers of volatile lighter elements are blown millions of miles into space, and the heavier elements collapse into a shrunken superdense sphere perhaps less than one one-thousandth of its original size.

Aside from being one of the most spectacular cosmic events (and ultimately frightening as it could happen near our own solar system), supernova explosions are interesting because of what they tell us about cosmic evolution. Some of the most exciting and perplexing creatures in astronomy's modern bestiary are thought to be spawned by supernova explosions: pulsars, black holes and other X-ray sources, expanding nebulas, and "runaway" stars speeding through the universe at speeds up to a million miles per hour. In addition, the outrushing energy from supernova explosions is thought to produce high-energy cosmic rays, radio emissions, and gravitational shock waves.

Cosmologists, those scientists concerned with the origin and evolution of the universe, also think a supernova explosion may be a miniature version of the "big bang" that created the entire universe and thus may duplicate on a smaller scale many of the same primordial conditions of creation. The heavy elements, dust, and gas spewed forth from an exploding star may provide the basic materials necessary to create new stars, new planets, and new life forms. Indeed, all living creatures—including humans—are made from elements ejected billions of years ago from a now dead and forgotten star. Some experts even suggest that the explosion of a nearby (relatively speaking, of course) star sometime within the past twenty million years may have flooded the earth with enough cosmic radiation to cause the mutation of certain species—and the extinction of others.

Because the supernova phenomenon is so central to much of today's astrophysics, astronomers would dearly love to observe an explosion first-hand with modern instruments. However, while statistics suggest that at least one supernova per century should be seen in our Milky Way galaxy, none has been recorded since 1604, when the last was observed by Johann Kepler. Of course, they may have occurred and we simply missed them. Supernovae tend to occur near the central plane of the galaxy, and the vast amounts of dust between earth and the galactic center may obscure our view. (Other stellar explosions, known as novae, are seen more frequently, with six already recorded in the twentieth century, one as recently as 1975. But as their name implies, novae are less violent and less interesting events, releasing only about one ten-thousandth the energy of a supernova, and remaining bright for a much shorter time—a few days or weeks, rather than months. The mass of the original star is much smaller in the nova, and less of the outer shell is expelled into space.)

As a result, astronomers have recently turned to ancient astronomical records to find accounts of supernovae observed in history. The hope, of course, is to find evidence linking the sudden brightening of a star in the past (or the sudden appearance of a new star, since many pre-explosion objects were completely invisible to the naked eye) with its remnant seen today as an expanding nebula, pulsar, or X-ray source. Ideally, these historians search for information that can describe in detail the star's position, brightness fluctuations, color, and length of visibility. By knowing a supernova's exact date and appearance, they hope to derive a standard for determining the expansion rate of the gas clouds. Moreover, they hope accurate dating can help explain assumptions about how cosmic debris is propelled through space.

Two of the more diligent record-hunters, F. Richard Stephenson and David H. Clark, have identified seven supernovae during the past two thousand years that remained visible for six months or more and that were recorded by ancient astronomers (primarily those in the Far East). The supernovae include one in the constellation Centaurus, seen in A.D. 185 by the Chinese; one in the constellation Scorpius, seen in A.D. 393, again by the Chinese; one in Lupus, seen in 1006 by Chinese, Japanese, European, and Arabic observers; two in Cassiopeia, the first seen in 1181 by Chinese and Japanese astronomers, the second in 1572 by Chinese, Korean, and European astronomers (this event was recorded by Tycho Brahe, the greatest of the pretelescope observers); and one in Ophiuchus, seen in 1604 by Chinese, Korean, and European observers. Understandably, the last two supernovae have been particularly well documented.

Still, the best-known supernova—and the first for which an age has been precisely determined—is the explosion associated with the Crab nebula and observed in the constellation Taurus on July 4, 1054, by Chinese astronomers. The "guest star," as it was called by the Chinese court astronomers (astrologers), outshone everything else in the night sky for nearly two years; and for a period of twenty-three days, when it was at maximum brightness, it could even be seen during daylight hours. The first historical connection between the Chinese "guest star" and the Crab was made in 1921 by Knut Lundmark of the University of Lund, who noted that the two objects might be the same. In the 1940s, Edwin Hubble of the Mount Wilson Observatory, by measuring the rate at which the visible filaments of the Crab were expanding, determined that the nebula must have been created within at least a century of the recorded Chinese observation. Further research by Virginia Trimble confirmed Hubble's theory and narrowed the date to within a few decades of 1054. More recently, radio studies of the Crab's pulsar show that its pulse frequency is lengthening because its period of rotation is decreasing with time. The rate of slowdown is consistent with an origin some nine hundred years ago.

One of the great mysteries of astronomy, however, is why people elsewhere did not see this extraordinary event nine centuries ago. No clear records have ever been found that astronomers, scholars, or even casual observers noted this brilliant star over Europe or the Near East. Most explanations for this lack of observation are unsatisfying. For example, some historians suggest that the Aristotelian concept of "perfection in the heavens" was so pervasive in medieval thought that no ecclesiastic (monks were the primary keepers of scientific records during that period) would have dreamed of noting an event that destroyed the apparent symmetry of the heavens. But since monks and scribes did note the appearance of comets and other obvious astral aberrations, the argument is not convincing. Nor is the suggestion that the Western world experienced inclement weather and a constant cloud cover throughout the entire period of the Crab's brightness. (Actually, some detective work by Kenneth Brecher and Elinor and Alfred Lieber in a medieval Arabic encyclopedia has uncovered a reference to an observation of an unusual star seen by Ibn Butlan, a Christian physician living in Constantinople around 1050. Brecher and the Liebers suspect that Butlan may have seen the Crab, for he associates the sudden appearance of a new star with an epidemic in the Near East recorded in 1054.)

Although Western literature provides little confirmation of the star seen by the Chinese, some new and unwritten evidence from the American Southwest may offer additional support to the connection between the Crab and the Chinese "guest star."

In 1954 William Miller, a photographer with the Hale Observatory and an amateur student of Southwest Indian rock art, was puzzling over two enigmatic pieces he had found in northern Arizona. One was a petroglyph, a design carved into the rock surface; the other was a pictograph, a design painted on rock. Both showed the same design: a crescent shape and a circle.

Astronomical motifs are common in Indian rock art, of course (see Figure 60), but this particular design combining a crescent moon and some bright globular object had not been seen—or at least noted—before (see Figure 61). An immediate interpretation of the design might be a lunar-planetary conjunction, similar to

Figure 60. *A wall of petroglyphs from Arizona's Painted Desert shows a variety of common themes in American Indian rock art: human and animal figures, hands, abstract designs, and astronomical symbols, including sunbursts, stars, and crescent moons.*

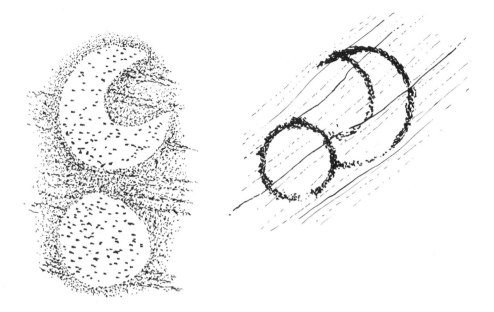

Figure 61. *Author's rendering of the two examples of rock art first associated with the Crab supernova by Miller.* Left: *A petroglyph from the canyon wall near Navajo Canyon, Arizona;* Right: *A pictograph from a cave near White Mesa, Arizona.*

what is seen on the flags of many Moslem countries. Such conjunctions are so common in nature one would expect the theme to appear often in Indian art. It does not. Thus Miller suspected the design might represent something more spectacular: a once-in-a lifetime event that profoundly affected the observer.

Following a suggestion by the British astronomer Fred Hoyle, Miller investigated the possibility that the crescent-and-circle motif might actually commemorate the appearance of a bright supernova—specifically, the supernova of 1054. An unusual astronomical coincidence made Miller's theory about the crescent and circle very plausible. Just before dawn on July 5, 1054, the crescent moon was in Taurus, placed so it was in close conjunction with the supernova. Moreover, since the moon moves through the star background approximately the width of its own diameter each hour, the conjunction would have been seen only on that morning and only in western North America.

For the Hohokam and Anasazi peoples watching the eastern horizon for sunrise on that morning, a star of truly extraordinary brightness would have appeared within 3 degrees of the crescent moon. The nearness of the event in time to the summer solstice—a moment of great importance to all North American Indian tribes—suggests strongly that some skywatcher would have recorded the phenomenon.

In the twenty-five years since Miller first noted the distinctive crescent-and-circle designs, over twenty additional examples have been found among the ruins and rock shelters of the American Southwest. The styles and media of these designs differ widely from site to site. Some are "pecked" or carved from stone; others are painted or drawn using a variety of materials. The shape and size of the two design elements, as well as their orientation to each other, also vary. Sometimes both elements are equal; sometimes one or the other is larger. The horns of the crescent moon sometimes point toward the star; other times they point away. Often the circle is really starlike, rayed like a child-drawn sunburst; sometimes the "star" is more a cross; but most often it is simply a dot or filled circle. Except for a few examples, the crescent and circle appear on rock walls that are covered with many other symbols, not all of them astronomical (see Figure 62).

Generally, the seekers of Crab-nebula evidence in Indian rock art have set three requirements: (1) The art must include a crescent moon in either waxing or waning phase appearing very near a circle, pit, or starlike form; (2) the art must be exposed to a northeastern horizon from which the Crab phenomenon might have been seen; and (3) the archaeological evidence at the site should be consistent with a possible habitation date of A.D. 1054.

In fact, almost all the twenty-odd examples of art have been found at sites where there is good evidence for human occupation at this period. Of course, it would be ideal if the art itself could be radiocarbon dated, a test that could be performed on those pictographs which have organic materials as their medium. However, at present this test cannot be easily done without destroying the art itself.

184

Figure 62. *Author's rendering of rock art possibly depicting the Crab supernova explosion. Top: A carved crescent and concentric circle from the Capitol Reef National Park, Utah;* middle left; *a painted crescent and star found near Scholle, New Mexico;* middle right: *petroglyph from cave wall near the village of the Great Kivas, northeast of Zuñi, New Mexico;* bottom: *pictograph from Chaco Canyon, New Mexico. For scale, the hand symbol is nearly life-size.*

It is interesting, too, that several of the crescent-and-circle designs have been found at apparent sun-watching sites in the ancient pueblos. For example, John C. Brandt of Goddard Space
Flight Center, who is perhaps the best known of the many astronomers studying Indian rock art, has investigated one pictograph associated with a sun-watching shrine in Chaco Canyon
(see Figure 63). This particular work is unusual, for it is the only
known crescent symbol in the Canyon complex. "Further, the star
is stylistically unlike the stellar representations typical of later
pueblo tradition," wrote Brandt in a *Technology Review* article. "It
seems more than likely that the Sun-priest whose duty it was to
observe sunrise daily was struck by the spectacular association of
the waning crescent moon and the bright supernova and recorded
it on the spot."

Perhaps more than any other aspect of archaeoastronomy, much
of the Crab nebula's rock record must be taken on faith. Even if
accurate dating could be achieved, it would not provide definite
proof that the art represents this particular event. Several experts
in Indian ethnology argue that the symbols could well refer to any

Figure 63. *The crescent-and-star design painted on a cave ceiling at Chaco
Canyon, New Mexico, is thought to represent the appearance of the Crab
supernova near the thin crescent moon on the morning of July 5, 1054. For
scale, the hand is nearly life-size.* Photo courtesy Von Del Chamberlain

number of phenomena, including close planetary-lunar conjunctions. Klaus Wellman, for example, has also noted that Indian artists using hallucinatory drugs often produced fantastic drawings incorporating otherwise familiar symbols in contexts totally unrelated to observed reality. Other experts in primitive art warn that the same symbols and designs often had very different meanings to different peoples, even those people with the same general cultural roots (see Figures 64, 65, and 66).

One of the most persistent critics of the rock-art theory has been Ho Peng-Yoke, who has written in *Vistas in Astronomy*, Volume 13:

> The motif of crescent with the horns pointing toward a star or stars has appealed . . . to many peoples over long periods. In the Turkish flag (horns pointing to the right), the star is said to be Ai Tarek which is variously the morning star, Saturn, or the Pleiades, according to the Koran. Of the other thirteen national flags displaying that or a similar motif, nine show the horns pointing to the right, one to the left, one upward, and two show more than one star (horns to the right); of these two with the horns to the right, one shows three stars (flag now obsolete), and the other five stars. The motif appeared on the badge of Richard I (the star is thought to be the Star of Bethlehem) and is now used by the Shriners (branch of the Masons). In none of these cases has the 1054 event been invoked to explain the juxtaposition of crescent moon and stars, so why should that event be invoked to explain the Arizona Indians' drawings?

Ho Peng-Yoke actually takes issue with the entire assumption that the Crab and the object of 1054 are connected, citing several discrepancies and contradictions in the original Chinese records.

Brandt and his colleague Ray Williamson, in their definitive paper published in a special issue of *Archaeoastronomy*, admit that the evidence remains inconclusive. They base their case on the supposition that the supernova was extremely bright, that the crescent moon appeared near the supernova at maximum brightness, and that other nearby planets were absent from the sky in July

Figure 64. *The star-and-crescent motif also appears in some Hopi Kachina headdresses, as in the illustration at the top left. In the figure at top right, star images only are seen, while in the bottom left one the astronomical symbol seems to be a sun. At the bottom right no astronomical symbols are present.* Illustrations courtesy Smithsonian Institution, National Anthropological Archives

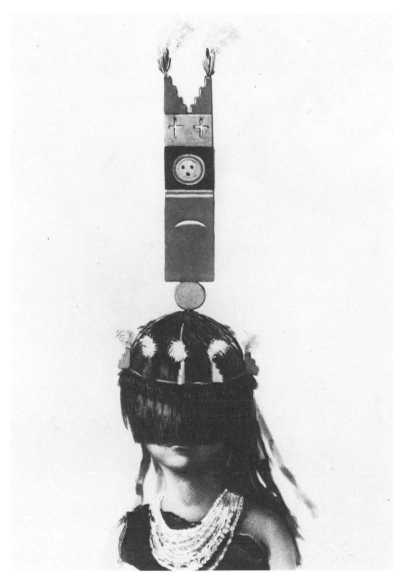

Figure 65. *Astronomical symbols appear in this ceremonial headdress worn by Zuñi Indians during the autumn rain dances that follow the appearance of the morning star* (Venus) *on the horizon. The sun (a circle with three dots) in the center apparently shines on the lunar crescent below, while the two four-pointed stars above represent Venus as both the morning and evening stars.* Illustration courtesy Smithsonian Institution, National Anthropological Archives

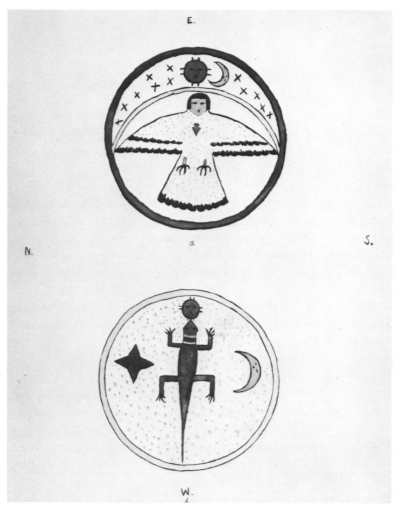

Figure 66. *Two sand paintings used by the Acoma Indians for ceremonies of curing clearly show the crescent-and-star motif.* Illustration courtesy Smithsonian Institution, National Anthropological Archives

1054. Given the relative rarity of other crescent symbols in rock art, and the propensity of preliterate men to notice and record unusual sky events, they think the likelihood of a Crab-nebula observation is great.

"Certainly the spectacular appearance of the supernova near the waning crescent moon was a sky event of major proportions and

190

one which must have filled the skywatcher with wonder," they write. "Whether that skywatcher was moved to record the event cannot yet be proved, but the rock art greatly strengthens the circumstantial case."

Unfortunately, even if the hypothesis that Amerindian astronomers recorded the Crab nebula explosion proves correct, the records they left behind are still little help in defining the accurate position, color, or brightness of the star, or how long it remained visible. For Stephenson and Clark, and others seeking historical confirmation of supernovae, the most useful records remain those found in the written chronicles of the Far East, medieval Europe, and the Arab world. As such, these types of observations are somewhat beyond the purview of this book, or rather beyond its main concern: the astronomy of preliterate peoples. Astronomy in the Far East after the Christian era and in Europe after about A.D. 1200 had become systematized and formalized, even if its goals in both societies often remained primarily astrological. However, more information about supernovae still may be found among the remains of other ancient peoples, and one unusual example should be noted.

In 1971, Brandt and two NASA colleagues, Stephen Maran and Theodore Stetcher, wrote a letter to the journal *Archaeology* seeking the assistance of archaeologists in identifying records of any bright supernovae observed from the southern hemisphere some five thousand to fifty thousand years earlier. Specifically they wanted evidence of a star explosion related to the nebula in the constellation Vela. This object is the closest nebula to earth and has a powerful X-ray source associated with a radio and optical pulsar at its center. Vela's expanding gas shell has been particularly difficult to date by standard astronomical techniques, although the apparent slowdown rate of the spinning pulsar suggests an origin between ten thousand and six thousand years ago, or around 4000 B.C., about the time the first civilizations emerged in the Near East.

Their unusual request was reprinted by *Time* magazine in March 1972, along with one of the pictures taken by Miller two decades earlier. Responses by *Time* readers about other examples of Indian rock art set Brandt and his colleagues off on a hunt for Crab-nebula connections in the Southwest. The original appeal for information on Vela was apparently forgotten, except by George Michanowsky.

A research consultant, explorer, and amateur archaeologist with a special interest in ancient languages, Michanowsky remembered an odd rock carving that he had found in the Bolivian Andes near the headwaters of the Amazon. The petroglyph showed four small circles joined by lines in the shape of a cross. Two of the circles were significantly larger than the others, and the largest had lines extending from it like the rays of the sun. Michanowsky interpreted the design as representing the so-called False Cross of the southern sky, as distinct from the well-known Southern Cross.

The smaller of the two largest circles, he felt, was the star Canopus and the larger of the two circles was a supernova. The area where the carving was found still had great significance to local Indians and served as the site for a major religious festival. However, no habitations were near the carving; thus, like much of Brandt's Southwestern rock art, it was impossible to date. At best the interpretation of the carving as a representation of the Vela supernova remained an inspired guess.

Michanowsky didn't stop there, however. Instead, he looked for similar records in the Old World. A supernova at the relatively close distance of Vela would have been the brightest object in the sky outside of the sun and moon. It would have been impossible to ignore. While the explosion would not have been visible from Northern Europe, it was possible, he felt, that observers in the southern Tigris-Euphrates Valley could have seen the bright star low on the southern horizon over the waters of the Persian Gulf.

A tantalizing line in the Old Testament further directed him toward Mesopotamia. In Job 9:9 there is a reference to the powers of the Lord "which maketh Arcturus [the Bear], Orion, and Pleiades, and the chambers of the south." In some versions of the Bible, this last phrase is translated: "the inner parts of the south." Michanowsky also discovered that the Italian astronomer Giovanni Schiaparelli (perhaps best known for describing the lines on Mars as "canali," or channels) had delved into the early history of astronomy and had become fascinated by this same phrase. Schiaparelli suggested the "chambers" or "inner parts" referred to a "secret" of the southern sky. Schiaparelli himself was puzzled by this reference, for he could not imagine any "secret"—or any object, for that matter—in the south which by its nature was more interesting than objects of the northern sky. Michanowsky became convinced that the "secret" was perhaps a transient event observed and recorded by skywatchers in an early Mesopotamian society, most likely the Sumerians. By the time the record was filtered through Babylonian texts, Hebrew translations, and finally into Latin versions of the Bible, the original concept had lost all meaning.

Michanowsky then turned to the original Middle Eastern records to find the origins of the "secret," which he assumed could have been a brilliant star seen low to the south. On Sumerian cuneiform tablets, he found a reference to a star named Mul Nun-ki sacred to Ea, god of the "Southern Sea," or Persian Gulf. The tablet contained a tale describing "the yoke star of the ocean," which was interpreted by Michanowsky to be a poetic convention for "a large star low on the southern horizon." More to the point, says Michanowsky, the star Mul Nun-ki was located in the Vela constellation, a sky region sacred to Nin-mah, the Sumerian goddess of life and protection and the primordial mother of the universe. When combined with other myths and symbolism in Sumerian records, including the persistent belief that the gods came from the sea to bring learning to men, Michanowsky felt certain that the people of the Tigris-Euphrates Valley had observed the supernova. Moreover, he argues that the experience may have been so profound that it actually formed the basis for their religious system, inspired the need for writing (or at least record-keeping), and touched off the rise of Sumerian civilization.

Of course, many archaeoastronomers—both professional and amateur—have discovered to their chagrin that drawing inferences from a mass of unrelated or tenuously connected data can be unwise. Myths and legends are never literal recountings of actual events, so almost any interpretation can be made to sound plausible. As mentioned earlier, even the more conservative assumptions of Brandt and his colleagues have been criticized by ethnologists who warn that the traditions of astronomical recording suggested by modern folklore may not have operated in ancient cultures, even when the modern peoples are the direct descendants of those societies. Similarly, myths and legends transmitted through several millennia almost always are transformed by the transmission. While Michanowsky's theory is an original, imaginative, and even plausible example of a supernova observation, the evidence remains, like that of the Southwestern rock art, purely circumstantial.

The problem of applying current myths and cultural traits to the task of determining the astronomical abilities of a people's pretechnical ancestors is best seen in the strange case of the Dogon people of West Africa and their relationship to the bright star Sirius.

194

Light
on the
Dark Continent

"It is characteristic of most colonial societies to seek justification of their colonization in historical precedent."
—Brian Fagan

In 1871 an eccentric thirty-four-year-old Swabian geologist named Karl Gottlieb Mauch, who had hiked alone through much of South Africa, stumbled into the ruins of a sprawling stone city in Rhodesia's Mapudzi Valley. Two high-walled buildings constructed of flat granite bricks laid in intricate herringbone and chevron patterns covered the top of a steep hill, and dozens of smaller stone structures were spread out in the valley below. The most prominent building was what he called "the Acropolis," a warren of curving walls and narrow passages perched at the edge of a ninety-foot precipice. An even larger structure, named by Mauch the "Elliptical Building," was located about six hundred yards below the Acropolis. This building had a roughly oval wall 250 feet in diameter and 830 feet in circumference. The walls were unmortared layers of rock, twenty to thirty feet high. They tapered slightly

195

from bottom to top and were neatly rounded at their ends. Three entrances in the outer wall opened onto a series of interior walls forming a maze of passages, hallways, and rooms. Two towers, both originally more than thirty feet high, stood at the southeast side of the main building.

Mauch returned to Germany to write a book about his discoveries. He was convinced that the ruins, and the rituals of the contemporary native groups living near the site, were Semitic in origin. In fact he believed the ruins were really those of the biblical Ophir, a legendary site from which King Hiram of Tyre brought gold for Solomon. The Acropolis was an imitation of King Solomon's temple on Mount Moriah, said Mauch, and the Elliptical Building was designed after the palace in which Bilquis, the Queen of Sheba, had resided during her visit to ancient Israel. Mauch further claimed the ruins were the home of Bilquis herself, who had imported Phoenician workmen to construct them.

In reality, the ruins "discovered" by Mauch were constructed by black Africans, using local technology, and within recent history. During the sixteenth and seventeenth centuries, this site was the capital of a widespread black empire that flourished in the area between the Zambesi and Limpopo rivers and was ruled by a king called the Monomotapa. By the time the Portuguese arrived in the late 1500s, the Monomotapa controlled all trade in gold and slaves between the interior and the Arab enclaves along the coast. By 1629 the Portuguese had replaced the Arabs and turned the Monomotapa into a puppet ruler. They then systematically destroyed the once powerful empire by intensifying the slave trade to the exclusion of all other activities. Eventually the Monomotapa was overthrown and his empire divided by other tribes. His glittering capital city—known as Zimbabwe—fell into disrepair. There is some evidence, however, that the city was rebuilt and reoccupied several times thereafter, perhaps the last time only a generation or so before Mauch found it in ruins.

Although Mauch's theories were pure fantasy, they did fit well with nineteenth-century misconceptions—and politics—concerning Africa. A host of other would-be historians and amateur archaeologists followed his lead and proposed that the builders of

Zimbabwe were Arabs, or Dravidians from southern India, or sea-farers from Malaysia. One explorer, Carl Peters, claimed Zim-babwe was the legendary "Land of Punt," even though ancient Egyptian records placed Punt somewhere near modern Somali on the Gulf of Aden. The best-known expression of these theories was H. Rider Haggard's enormously popular (and influential) adven-ture novel *King Solomon's Mines* (1885).

Obviously, most European colonizers of Africa could not con-ceive of native black peoples with the intelligence, skill, or organi-zation necessary to build such complicated and elegant structures. Even the more sympathetic Europeans must have been reluctant, as Professor Jacques Barzun has suggested, to believe that a civi-lized people could relapse into ignorance and barbarism. Thus the spate of romantic novels and modern myths about ancient, pre-African, white civilizations that had flourished and failed centuries earlier served to answer the questions about the ruins and the traces of technology among the native peoples. The attitude was similar to the nineteenth-century American attempt to attribute the Mississippi Valley mounds to anyone other than the Indians. In the end, perhaps, the theories also helped the Europeans ratio-nalize their takeover of the continent by hinting that these lands had actually belonged to a higher (and most likely white) culture originating outside of Africa.

Even today, Africa below the Sahara is a remote, mysterious, and somewhat baffling place for Westerners. To the Europeans before the mid-nineteenth century, however, it was truly *terra in-cognita:* a hostile, pestilent, and dangerous world filled with canni-balistic savages and ferocious animals. Africa was the dark continent, unilluminated by knowledge or reason.

The earliest European arrivals, primarily Portuguese military men, ignored any evidence that the black cultures of West Africa had developed sophisticated societies, systems of government, and traditions of art equal to those of the Egyptians and Sumerians. Al-though limited by the considerable climatic and geographical barriers of the continent, these cultures demonstrated remarkable technical and organizational skills. Agriculture, ceramics, metal-working, and textiles particularly had reached extraordinarily high levels of development. Yet within only a few decades, relentless and ruthless exploitation by the European colonists had reduced the existing African cultures almost to the levels of barbarism presupposed by the conquerors.

As late as 1977, Patrick Moore and Pete Collins, describing science in South Africa, could write that there was "little in the way of true astronomy before the arrival of the first Europeans." And they cite several standard myths to underline the simplicity of astronomical knowledge among the southern tribal groups. For example, one tribe's view of the universe described the earth as a clay bowl carried on the back of a giant tortoise swimming in the vast sea of eternity. When the tortoise dipped in one direction or the other, this blue surface of the sea could be seen and the sun would shine into the bowl. As long as the tortoise swam upright, the sun remained below the edge of the bowl and only darkness could be seen above.

To the pygmies of the rain forest, Khonuum was the supreme god of the sky, and each night when the sun died he collected its shattered pieces (the stars) in a sack and mended the broken sun so it could reappear the next morning.

For the Zambezi, the moon was once a pale and unshining globe. Jealous of the sun's brilliance, the moon waited until the sun went on the other side of the earth and then stole some of his fire. Angered by this theft, the sun threw mud on the moon's face, so that it was covered with dark patches. Filled with hate and a desire for vengeance, the moon waits for a rare chance to catch the sun off guard. When this occurs, the moon spatters the sun with mud, so that the sun may stop shining for several hours, causing the entire world below to be filled with fear and dread.

Although certainly imaginative, these views of heaven and earth do not reveal any real astronomical knowledge. On the other hand, there is evidence that several African groups, even in their folklore, had developed more sophisticated cosmologies. For example, one South African tribe perceived that the earth was in motion around the sun—a concept found rarely in ancient or preliterate societies. Moore and Collins themselves quote an old song: "I shall worship you and go around you, just as the earth worships the sun."

Also, the cave paintings of ancient African cultures, as well as the decorative and ritualistic art of many contemporary peoples, show what appear to be representations of the five naked-eye planets revolving around the sun. The symbols used by tribes throughout Africa to represent celestial objects include a crossed circle for the sun, a circle with a dot for the full moon (the full phase represented love; the crescent, tragedy), and the swastika, one of the most universal symbols, for light. Many of the familiar star groups known to European and American cultures also appear in different African folk traditions. For example, among South African tribes, Orion's belt is the "three healers"; the Pleiades are the "Ploughing Stars" (suggesting a calendric use that signaled planting time); the Milky Way is the "Starry River"; and the Southern Cross is the "Tree of Life" as well as an important guidepost for travelers.

"Folklore of this kind is reasonably well authenticated," write Moore and Collins, "but it cannot be said to be rich, and there was no overall pattern comparable with that in many other countries."

There are two problems with that statement. First, a thorough understanding of the cultural heritage of Africa is only now emerging. Second, there has been a tendency to lump all the black peoples of Africa into a single cultural mold, when, in fact, the racial, ethnic, and tribal differences of Africa are many and great. Shaped by distinct climatic and environmental conditions, several widely separated groups developed complex systems for organizing their societies—and their view of the heavens.

For example, the Mursi people, herders and cultivators living in southwest Ethiopia, have developed a lunar calendar linked to the rains, planting, and harvests. In this remote and virtually isolated country, the first rains begin in March and planting follows immediately thereafter. The resulting and somewhat meager crops are harvested in June or July. But if the rains are poor or the crops have failed, the Mursi often have a second chance. A brief second flooding of the riverbed occurs in October or November and leaves the flat lands immediately beside the river receptive to seeding.

David Turton and Clive Ruggles found that the Mursi measure a period between the onset of the rains to the flooding and to the rains again as one year, or *bergu*. The "age of the bergu," or time of the year, is determined by watching the intervals between successive new moons and numbering these periods in sequence, 1 through 12. Each "month" takes the name of the special activity—planting, hunting, harvesting, dancing—that normally occurs in that particular period. Of course, like all lunar calendars, the Mursi months quickly go out of synchronization with the actual solar year. Without altering the system somehow, the "bergu of planting" would soon occur in the "bergu of harvesting." Thus, every third year the Mursi add a thirteenth, or leap, month to bring the lunar calendar back in line with the solar year.

While this is a standard procedure with most lunar systems, the odd thing is that no two Mursi can (or will!) ever give the same answer to questions about the "age of the bergu." When interviewed they are both very casual and very secretive about the actual date, usually suggesting that the man who knows the bergu "lives far away" or "has just died." There is little argument among the Mursi over this inexactitude. In fact, as Turton and Ruggles found, the Mursi "agree to disagree" about the date.

More strange, the Mursi also use other, fairly accurate astronomical observations to determine their solar year. The rising points of certain stars on the eastern horizon as well as the changing positions of certain groups of stars in the nighttime sky provide an indication of when the intercalation of a month is necessary to bring the solar and lunar calendars into line. Obviously, the same observations could settle any disagreements about the "age of the bergu"; but, by tacit agreement, the disagreement remains. Some anthropologists think this stubborn refusal to agree on agricultural dates persists because each farmer thereby feels he has some advantage over his neighbors by maintaining a private calendar. Or perhaps it is simply a case of what Catherine Callaghan notes as the tendency among preliterate agricultural societies to employ two calendars: a common one regulating agricultural affairs and an elite calendar for determining rituals. Although by no means scientific astronomy (and some might argue that it isn't even very practical astronomy), the Mursi demonstrate again that even very primitive societies may make direct use of simple observations to regulate their lives and to develop complex social customs.

200

One of the more complicated social structures is the so-called Gada system followed by the Borana branch of the Oromo, or Galla, people of southern Ethiopia. This large Cushitic-speaking tribe spread throughout most of northeastern Africa during the sixteenth century, largely because of the unusual demands imposed upon them by their own Gada system. The Gada system both established a hierarchical class ranking and then determined that these different classes would succeed one another approximately every eight years. At that time the different groups would all assume new and specific military, economic, political, or ritual responsibilities in an elaborate power-transfer ceremony.

Under the traditional rules, however, before the top, or Gada, class could assume power, its members were required to wage war against another neighboring community or tribe. But the attacked community had to be one which no Gada ancestors had ever raided before. Thus by its constant demand to find new opponents and new lands to conquer, the Gada system pushed these Cushitic peoples outward to cover half of Ethiopia and parts of northern Kenya. Today the movement of classes through the Gada system no longer requires open warfare, but the system is no less complex. Classes and castes are highly structured, and procedures for upward mobility are codified in a form and schedule related to an equally complex astronomical calendar.

"Borana time reckoning is unique in eastern Africa and has been recorded in very few cultures in the history of mankind," according to A. Legesse, an expert on Gada. This system consists of a permutation calendar based solely on a 29.5-day lunar cycle with no relation to the sun. Careful observation of certain clusters of stars appearing near each month's full moon allows adjustment of the calendar—through the addition of appropriate days—to make the lunar year correspond closely with the tropical year.

While the existence of the Gada calendar, and its need for the precise observations of stars, gives evidence enough for a non-European astronomy in Africa, recent archaeological discoveries in northern Kenya suggest that the astronomy underlying the Gada calendar actually may have been developed more than two thousand years ago.

Namoratunga is a large cemetery and rock-art site along the Kerio River south of Lake Turkana in northwest Kenya. The graves of Namoratunga are surrounded by huge standing stones engraved with cattle brands in the style of similar burial sites of the Cushitic peoples of southern Ethiopia. B. M. Lynch and L. H. Robbins of Michigan State University have found, through radiocarbon dating of other artifacts at the site, that the graveyard was first used around 300 B.C. In the language of the local tribe, *Namoratunga* means "stone people." The tribe believes an evil spirit who once dwelt here had the power to turn to stone anyone who mocked him. Because the standing stones of Namoratunga are so strikingly different from other graves in this area, the archaeologists inquired of tribal elders about any related sites in the vicinity. They were directed to a location some 210 kilometers north, on the western shore of the lake. This site was located on the slope of a mountain overlooking the lake basin, thus providing an unobstructed view of the entire eastern horizon. Central Island, a very prominent feature in Lake Turkana, was due east of the site.

Lynch and Robbins called this site Namoratunga II, for they found it remarkably similar to the first graveyard and belonging to the same period. At least one grave was marked by an upright slab, and other standing stones bore the faint remains of cattle-brand symbols. There was one significant difference, however. At Namoratunga I, the standing stones only circled the graves; here, nineteen stone columns were arranged in rows unrelated to the graves. Moreover, the alignment of the upright basalt slabs seemed deliberate, not simply random.

Because the present-day eastern Cushites, of whom the Boranas are a part, maintain a calendar which uses the risings of seven stars or constellations in conjunction with phases of the moon to calculate a 12-month, 354-day year, the archaeologists decided to test for any correlation between the stone arrangements and the stars.

The Cushite calendar uses the stars Aldebaran, Bellatrix, Saiph (Kappa Orinis), and Sirius, plus the star groups Pleiades, Central Orion, and Triangulum. During half the year, the Cushites identify months with the rising of the stars in conjunction with the new moon, with each star or group appearing successively in the order—Triangulum, Pleiades, Aldebaran, Bellatrix, Central Orion, Saiph, and Sirius. During the second half of the year, only Triangulum is used, beginning when it rises in conjunction with the full moon. The following months are then identified by Triangulum's relationship to the phases of the waning moon.

Using 300 B.C. as a probable date for construction, alignments between certain pairs of stones were checked against the azimuths of the rising stars during that epoch. Lynch and Robbins also assumed that sightings would have most likely been made using a western stone as a backsight and an eastern stone as a foresight. Since most stones were set in the ground at angles (apparently still a traditional technique of the Cushites), they used only the tops of the stones as sight markers. Twelve alignments were found between two, and sometimes three, stones and all seven stars or constellations. In only three cases was the alignment more than one degree off the azimuth. In addition, they found all but one of the stones in the complex were used as sight markers in some alignment. (Since that one stone was only fifteen centimeters above the ground surface, they assumed it would have been of little value as a line-of-sight marker.) Although Lynch and Robbins did not test the theory, it seems reasonable to assume that Central Island, the only prominent horizon feature visible from the site, also might have been used as a distant foresight for some observations.

To test their alignment theory, Lynch and Robbins used the present-day azimuths of the star-rise points (some had changed nearly 12 degrees due to precession) and found only four star azimuths matched with stone alignments. One was Sirius, whose present azimuth is little different from its azimuth in 300 B.C.

In an article for *Science*, Lynch and Robbins wrote that "the archaeoastronomical information described for Namoratunga II adds significantly to the growing body of evidence attesting to the complexity of prehistoric cultural developments in sub-Saharan Africa. It strongly suggests that an accurate and complex calendar system based on astronomical reckoning had been developed by the first millennium B.C. in eastern Africa."

Archaeoastronomers hope that other megalithic sites will be found in Africa. However, for the present, the tracing of cultural traditions remains dependent primarily on the folklore and oral histories of contemporary peoples. As has been noted elsewhere, deriving scientific information from myth can be a risky enterprise. Indeed, one of the most controversial issues in archaeoastronomy is the case of the Dogon people and the myth of Sirius's invisible companion.

The Dogon people live in the Republic of Mali, once known as the French West Sudan, not far from the fabled, if now somewhat seedy, city of Timbuktu. Their homeland is thus close to one of the great desert caravan routes stretching from West Africa through the southern Sahara to Egypt. According to Colin Turnbull in his survey of African civilization, *Man in Africa:*

> This is an area in which an extremely early and indigenous development of agriculture may have taken place . . . and the Dogon may represent a people who stayed in their ancestral homeland while others pressed outward in search of better arable land. The Dogon are a people with a long tradition of cultivation, to whom it comes easily despite the difficulties of their environment, and who have developed an agricultural system that relieves them of much of the worry that faces others. They have a long and peaceful history of settlement, and are not overly concerned with economic factors. Content with the stability and subsistence level that they have achieved, the Dogon turn their energies to a consideration of the source of being and have developed an elaborate artistic expression of their belief in sculpture, music, dance and poetry.
>
> Where others, even their immediate neighbors and kin, concentrate on the problems of life and living, the Dogon concentrate on another realm of being altogether. The social system runs parallel to the world of belief as expressed in Dogon myth. But Dogon myth is itself a dual system—an esoteric and exoteric body united in a complex system of symbols that the living society and social organization represent. . . . Their unity springs from the belief that God and earth were lovers and that man is the seed of the universe.

Understandably, perhaps, a people with such a philosophical interrelationship with the spiritual and physical world also has an elaborate set of myths concerned with the sky: a cosmology that describes both the creation of man and his observable environment. It is most strange, then, that an apparently major portion of their culture and religion is based on the existence of supposedly *unobservable* celestial bodies and motions.

Until the early part of this century, the Dogon apparently had little contact with European civilization except for the occasional trader or missionary. In the 1940s, however, two French anthropologists, Marcel Griaule and Germaine Dieterlen, came to study and live with the Dogon for varying periods extending over the next twenty years. The two researchers became so familiar and so trusted by the tribe that Griaule was finally allowed to hear the secrets of Dogon cosmology. Through both oral interviews and complex diagrams drawn in the sand, the Dogon chiefs and priests described how the earth and all the planets rotate around the sun in elliptical orbits, how Jupiter has four satellites, and Saturn has rings. Since the first fact is not immediately apparent to a non-mathematical culture, and because neither the Jovian planets nor the rings of Saturn can generally be seen without telescopes, such beliefs were extraordinary. But even more startling was the Dogon belief about Sirius, the central and most sacred celestial body in their universe.

The Dogon informants told Griaule and Dieterlen that Sirius was really two stars, not one. A dark unseen companion, the tiny star Digitaria, circled Sirius in an elliptical orbit with a period of fifty years. Known to the Dogon as "the beginning and end of all things" Digitaria was both the smallest and heaviest object in the sky. Consisting of a metal called sagala, which was described as a little brighter than iron, the star was so heavy "that all earthly beings combined cannot lift it, for the star weighs the equivalent of all seeds, or of all the iron on earth, although it is the size of a stretched ox-skin or a mortar."

The odd thing, of course, is that Sirius does have a small dark companion, better known to astronomers as Sirius B. In fact, Sirius B was first discovered in 1862 by Alvin Clark of Cambridge, Massachusetts, using one of the fine refractor telescopes he and his son crafted for many nineteenth-century observatories. An understanding of Sirius B's very peculiar properties awaited the development of larger telescopes—and the theory of general relativity—in the 1920s. By 1925 astronomers realized that Sirius B was a white dwarf—that is, an old and dying star whose outer atmosphere of lighter elements had been blown away to leave an extremely compact core of heavy elements. In fact, while Sirius B is perhaps a hundred times smaller than our sun, it has a density nearly a million times greater. This small, dark, heavy companion orbits Sirius with a period of exactly 49.9 years.

In his definitive study of the many "Sirius enigmas" appearing in *Astronomy of the Ancients*, Kenneth Brecher has reviewed the various theories suggested to explain the Dogon's remarkable prescience about Sirius. For example, it is possible, but extremely unlikely, that Griaule simply falsified—or at least misinterpreted—the story of the Dogon priests. However, Griaule (now dead) and Dieterlen were known as sober, serious, and conscientious scholars who thoroughly understood their subjects and whose resultant publications could hardly be considered sensationalist tracts for public consumption. They had no reason to create a false myth. Some historians have suggested the information about Sirius came along the trade route from Egypt several millennia earlier. But this in turn implies that the Egyptians held similar beliefs; and while Sirius certainly was central to the Egyptian calendar and astrology, no reference to its "companion star" exists in Egyptian texts. In fact, as Brecher points out, the Egyptians left no quantitative descriptions of the star, never describing even its color, position, or other physical properties.

The best-known and, to some, most popular explanation for the Dogon's knowledge of Sirius comes from Robert Temple, who spent nearly a decade researching the legend. He proposed that the knowledge of Sirius B, as well as the other bits of astronomical knowledge attributed to the Dogon, was brought to them by extraterrestrial visitors. His book *The Sirius Mystery* suggests that earth may have been visited by travelers from a planet in the Sirius system who imparted stellar lore to their hosts here. The theory, while interesting, unfortunately crumbles under even the most gentle prodding. Why, one asks, would these space visitors provide only partial information? They would have surely told the Dogons—or whoever else had listened—that Jupiter had not only four, but at least eleven other moons, most of them invisible from earth even in large telescopes and observable only from close-encounter spacecraft flybys in the late 1970s. And why would they have stopped with describing Saturn's rings alone, when both Jupiter and Uranus have similar features—again observed for the first time only in the 1970s?

A more logical explanation for the Dogon's remarkable cosmology seems one proposed independently by Brecher and by Carl Sagan. They think the Dogon heard the story of Sirius B sometime before Griaule and Dieterlen arrived in West Africa, and somehow the "modern" science became incorporated into the tribe's "ancient" myths.

The physics of Sirius B was certainly well known by 1925, for Brecher found several reports in the popular press of the time. Indeed, he feels the strange nature of the white dwarf may have become a "hot topic" for amateur science buffs of the 1920s, just as black holes have become today.

From his research, Brecher has devised a delightful scenario, given in *Astronomy of the Ancients:*

> Some Jesuit priest reads about it in *Le Monde,* and then goes to Mali long before Griaule and Dieterlen. "Tell us your myths," say the Dogon. "Do you see that star?" replies the priest. "It is actually two stars and the invisible star is the heaviest thing there is!" The Dogon promptly incorporate this information into their culture. And when the two anthropologists are told the secrets of the Dogon, all they get is a cross-cultural translation.

The Dogon, incidentally, also think there is a third star—also invisible—associated with Sirius A and B. Brecher found that the "third-star theory" also had wide circulation in the 1920s. But attempts by astronomers to locate such a triple companion then, as well as in the early 1970s, were unsuccessful. Recent research, using techniques refined considerably since the 1920s, indicates that the perturbations in the orbit of Sirius A which gave rise to the hunt for companion stars in the first place are probably well accounted for by a single companion—Sirius B. All of which suggests to Brecher that the Dogon "myth" became locked in place about fifty years ago and they (or any visiting missionaries) did not keep up with the scientific literature.

The Dogon's other "unobservable phenomena"—the moons of Jupiter, rings of Saturn, and the ellipticity of the planetary orbits— also might have been introduced by outside, but earthly, visitors. After all, the four moons of Jupiter have been known since Galileo's observations in 1610. And, sitting along a busy caravan route, the Dogons could well have received such information, albeit spotty and piecemeal, from the outside world.

Many anthropologists argue it is impossible for such foreign elements to penetrate to the very core of a society's religion. Yet examples of wholesale cultural corruption abound; most notable are the cargo cults of the South Pacific which worship the "big birds" that bring gifts from the sky. Of course, these people are waiting in vain for the return of the World War II military transports that once brought material goods to the islands. And the Cuna Indians of Panama now incorporate advertising slogans and television test patterns into the designs of their *molas,* those intricate and colorful panels created from layers of cut cloth. Moreover, the supposed "ancient art" of the mola itself dates only from the Conquest and the introduction of dyed cloth, steel needles, and European appliqué techniques.

Eddy has noted the reverse of the Dogon experience among the Plains Indians, where an astronomical tradition was obliterated almost overnight. "Practical solstice markings, using sun and stars, had been in Plains lore for 2,000 years," he wrote in *Technology Review.* "Yet, by the time of the last century, it had all been forgotten. With the coming of the white man and his horse and his calendar and his ways, a pure and primitive natural astronomy may have been knowledge no longer needed."

The controversy over the Dogons' Sirius myth will surely persist. The arguments of Temple and the other advocates of the extraterrestrial connection are simply too attractive and too romantic to be affected by the more prosaic and pragmatic suggestion that the Dogon were influenced by modern mass communications. A sadder aspect of the Dogon controversy is that it detracts from the more promising archaeoastronomical research just now beginning in Africa. The evidence from Namoratunga, for instance, suggests that a real and rich tradition of astronomical awareness existed in prehistoric time. The goal of many scientists is to understand this hitherto ignored and forgotten area of human intellectual development.

Stars
in the East

"A compass may go wrong, the stars never."
—Tongan saying

Among the earliest known records of an eclipse is this inscription on a bone found near An-yang in northeast China: "Three flames ate the sun and a great star was seen." Or at least historians suspect it is an eclipse record. Certainly this is an accurate, albeit thumbnail, description of the sudden flaring out of the sun's glowing outer atmosphere, or corona, from around the darkened solar disk at the moment of totality and the almost miraculous appearance of an otherwise invisible planet (probably Venus) close to the sun.

This inscription and others equally enigmatic were carved into tortoise shells and animal bones by the priests of the Shang dynasty sometime between 1500 and 1000 B.C. "To seek supernatural advice by divination," writes anthropologist Hung-hsiang Chou in *Scientific American*, "called for the preparation of 'oracle bones,' commonly the shoulder blade of an ox or the plastron (bottom

210

shell) of a turtle in a special way. . . . After the oracle bone had been prepared, the diviner engraved his question on its smooth surface, exposed the bone (or shell) to heat, and judged from the cracks that appeared whether the oracular answers to his questions were favorable or otherwise." Thousands of these engraved bones have been recovered from Shang sites since 1899 and they provide almost the only record of the intellectual pursuits, including astronomy, of these ancient people. (An inestimable amount of history may have been literally consumed by the modern Chinese, for there once was an active trade in "dragon bones" among apothecaries, who ground them into powders for sale as aphrodisiacs and pain killers.)

In addition to scattered solar and lunar eclipse references, the Shang bones also include occasional mentions of stars and novae, plus one or two tantalizing hints that the Shang divided the year into seasons by measuring the changing length of a shadow cast by a simple sun-stick or gnomon (see Figure 67). The record is by no means clear, for none of the bones can be dated except in a most general way. For example, the eclipse described at the beginning of this chapter could have been one in 1370, 1330, 1304, 1230, or 1200 B.C.—all times when totality occurred over or near the site where the bone was found.*

A small fraction of the bones apparently also served calendrical purposes. And from them we know that the Shang maintained a lunar calendar with a repeating cycle of sixty days —that is, six weeks, each with ten days. Because of its lunar nature, the Shang calendar needed regular adjustment, and a "leap month" was added every third year. Chou suggests the sixty-day cycle may have evolved from the Babylonian zodiac. He also notes it "survives in China today, although its main application is in counting a cycle of 60 years rather than 60 days."

---

* As reported in *Sky and Telescope*, the earliest eclipse whose date is unambiguous was recorded on a tablet found at the ancient city of Ugarit in northwest Syria: "The day of the new moon in the month of Hiyar [April–May] was put to shame. The sun went down in the daytime with Rashap [Mars] in attendance. This means the overlord will be attacked by his vassals." Since the town was destroyed in 1200 B.C., the inscription most likely refers to the eclipse of May 3, 1375 B.C., which could be seen from this site.

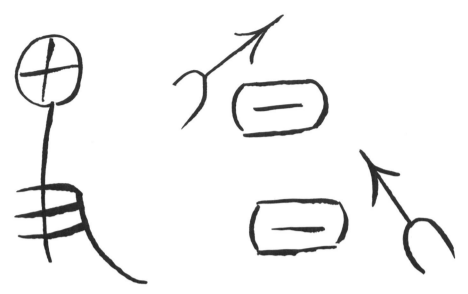

Figure 67. *Shang characters carved into so-called oracle bones show ancient astronomical symbols, including what may be a record of the oldest known astronomical instrument. Left:* The *symbol showing a hand holding a stick with the sun behind it may represent a simple vertical pole used to measure the changing length of the shadow and thus determine the solstices and the division of the year. The symbols at right are thought to be representations of the sun with the shadow of the human figure shown at different angles during the year.*

The Shang dynasty was succeeded in about 1100 B.C. by the Chou dynasty, which ruled until approximately the time of Confucius (551–479 B.C.). In fact, the *Spring and Autumn Annals,* edited by Confucius, contain over thirty observations of solar eclipses occurring between 720 and 481 B.C. It is the earliest continuous series of eclipses recorded by any civilization.

The Chou peoples also may have followed a practice widespread through much of the world, that of incorporating astronomical orientations into their architecture. Preliminary research by Vance Russell Tiede using aerial photography has shown that a significant number of pyramids in the Wei River valley of Shensi province have some celestial orientation. Hundreds of *lings,* or pyramidal burial tombs, dating from the early Chou period to the T'ang dynasty of A.D. 900 dot this valley. A remarkable number are oriented to the cardinal points within less than one degree. Tiede also found two, both dating from the later Han period (200 B.C. to A.D. 200), that appear to have their southeast sides aligned with the rising point of Sirius.

Sadly, almost nothing is known of Chinese history or science before the Shang period, or, for that matter, much after it until about 200 B.C. Again the problem lies with an unwise action by a succeeding generation. In 213 B.C., Ch'in Shih-huang (Qin Shihuangdi), the unifier of China and its first emperor, ordered the "burning of the books" to wipe out any memory of the former warlords. As the libraries, archives, and government documents disappeared in the flames, so did most links to China's prehistoric past.

Despite the loss of these records, it seems certain that astronomy emerged very early among the preliterate societies of China; and perhaps as early as 1000 B.C. it had become formalized into a rigid and complex astrological system.

Renewed interest by the People's Republic government in the lost history of ancient China has spurred extensive archaeological excavations, particularly at Shang sites. The possibility is great that future discoveries may offer new information on and insights into technology and science in China prior to 1500 B.C.

By contrast, the history of Chinese astronomy after 200 B.C. is extraordinarily well documented, with the records continuing almost uninterrupted into modern times. Most of the science is contained within dynastic histories. These great ponderous, multivolumed works were compiled faithfully by the Bureau of Historiography, whose civil servants survived the rise and fall of each dynasty to chronicle the fortunes of the former rulers. Most dynasty records included a long section on astronomy, listing all the various observations made throughout a particular reign and any astrological links to the fortunes of a ruling family.

Because such a full—and often poetic—record of astronomical observations exists, it is questionable whether a discussion of Chinese, or Japanese and Korean, astronomy is really appropriate for a book devoted primarily to the science of preliterate and pretechnical societies. However, a brief summary may be helpful in comparing Oriental cosmologies with those of the West.

213

From the Han dynasty (202 B.C.–A.D. 220) on, Chinese astronomers observed and recorded almost everything in the sky visible to the naked eye: lunar and solar eclipses, comets, novae, conjunctions of planets, occultations of the stars and planets by the moon, daylight sightings of Venus, the northern lights, and even sunspots. By the time of Christ, the right ascension (azimuth) of the sun was noted for all solar eclipses and theoretical books on astronomy had been written.

The great circle of the sky, or celestial equator, was divided into twenty-eight sections of arc called "mansions"; these represented the original political divisions of earthly China but also served as the right ascension element in their celestial coordinate system. The entire sky was further subdivided into approximately 250 asterisms, or small constellations of approximately a half-dozen stars.

Western astronomers recognize few familiar patterns in the Chinese zodiac. "The number of cases in which any parallelism of symbolic nomenclature can be made out is remarkably small," Joseph Needham noted in his definitive history. "And the same groups of stars were not seen in the same patterns."

There are some exceptions, of course: The Chinese, like ourselves, associated Orion with a human figure; and the popular name for Ursa Major was Pei Tou, or the Northern Dipper. Generally, the Chinese were more mundane than Western astronomers in their celestial symbolism, eschewing mythological heroes and gods for earthly personages and affairs. Thus, their stars and star groups had names like Emperor, Crown Prince, and Minister of Works; Celestial Temple, Outer Kitchen, and Guest House; Celestial Coin, Weeping Star, and Father-in-Law. Most historians cite the lack of similarity with Western asterisms as proof of Chinese astronomy's independent development.

The role of astronomy eventually became so important in China that a special subdepartment was established in the Ministry of State Sacrifices by the Han dynasty, presumably to assure that all celestial signs favored the time for making offerings. Later the Astronomical Bureau became a full-fledged government agency charged with observing the heavens and interpreting the celestial portents, as well as maintaining the exact time and calendar for the court. Imperial observatories were constructed in each of the various dynastic capitals; and when a dynasty fell and the capital moved, a new observatory was constructed in the new imperial city. Astronomers ranked near the top of the civil-service roster, having direct access to the emperor and being protected and shielded from the other bureaucrats. Indeed, during the T'ang dynasty, the astronomers were shielded from nearly everyone and forced to pledge complete secrecy about their work.

Concern with the heavenly events and their portents was not confined to the emperor alone, however. The influence of astrology was pervasive throughout Chinese society, affecting almost every aspect of daily life. Incredibly, this formalized, stylized, institutionalized astronomy, an exercise performed exclusively for the purpose of making prognostications, continued unchanged into the late nineteenth century and only disappeared completely (in official life, at least) after the death of the Manchu Empress Dowager Tzu Hsi in 1908.

Recorded history begins much later in Korea and Japan; and in both countries astronomy was strongly influenced by the Chinese. Around 50 B.C. the various independent and warlike tribes of the Korean peninsula were amalgamated into three kingdoms—Silla, Paekche, and Koguryo—under the watchful and protective eye of the Chinese. Chinese script was also introduced about this time, and by A.D. 400 its use had become widespread.

Korean astronomy shared the methodology and philosophy of the Chinese system and rapidly became formalized as part of the government bureaucracy. The official histories of Korean rulers also followed the style of their Chinese mentors, and long lists of astronomical observations—and their astrological implications—appeared in official court records.

In a book dealing primarily with the often frustrating and inconclusive attempts to find astronomical orientations in partial ruins, one early Korean building is worth mentioning, if only because it may be one of the world's oldest surviving structures known with certainty to have been an observatory.

This odd, bottle-shaped, stone tower is located about 210 miles southwest of Seoul in Kyongju, the former capital of the Silla kingdom. Erected in A.D. 657 during the reign of Queen Singdok, this *Cheom-seong-dae*, or "platform for observing the stars," is 30.5

feet high, tapering upward from a 17-foot-diameter base to a 10.5-foot-diameter neck like a giant version of an old-fashioned cream container (see Figure 68). The tower is constructed of 366 curved granite blocks, each 2.5 feet long and 1 foot high. The number of stones just happens to coincide with the number of days in a leap year. Moreover, the stones are piled in twenty-eight courses, corresponding with the number of "mansions" in the traditional Chinese "zodiac." The tower has an opening in its south wall through which astronomers apparently entered and then climbed to the top platform to make observations.

The Chinese system of writing came into general use in Japan around A.D. 400, but another three centuries passed before reliable histories appeared. Because only one imperial family has ruled Japan from the beginning of recorded time to the present day, no Chinese-style "dynastic chronologies" were compiled. References to astronomy are therefore scattered throughout many private histories, personal diaries, and academic archives. When collated and coordinated, however, these records show a well-developed system of astronomy beginning in the mid-seventh century. Like Korean astronomers, however, the Japanese were strongly influenced by Chinese methodology and philosophy; in other words, their goals were primarily astrological.

The first Japanese observatory also dates from this period. In A.D. 608 a priest named Mim was sent to study astronomy and Buddhism in China. He returned home trained to observe and interpret unusual celestial events. With the support of the emperor Temmu, Mim opened Japan's first observatory at Asuka on February 5, 675, and at the same time founded the Imperial Department of Astronomy. (The staff consisted of a president, two vice-presidents, two secretaries, six chief astrologers, one head of the calendar division with ten aides, one head of the observing division with ten aides, and two heads of the timekeeping division with twenty aides—double shift, perhaps—plus twenty messengers and two janitors.)

The observatory has been abandoned for more than a thousand years, and Asuka itself is a ghost town. All that remains are two carved megaliths possibly related to the original site. One stone

Figure 68. *The seventh-century observing tower, or* Cheom-seong-dae, *is one of the oldest stone structures in South Korea and one of the world's oldest known observatories.* Photo from the Korean Information Office, courtesy *Sky and Telescope*

has a series of grooves cut along its upper surface that supposedly point toward the solstice and equinox sunsets. The second megalith, estimated as weighing 950 tons, has two deep and sharply cut holes in its upper surface. It has been described as having legendary powers and mythical origins, but the markings may simply have been foundation holes for an observing tower. The most outstanding achievement—or, at least, most memorable—of the Asuka observatory was its discovery in October 684 of what later would be called Halley's Comet.

The general use of Chinese script by both Korean and Japanese astronomers has made the records of all three countries very accessible to historians and scientists. More important, after A.D. 1000 essentially the same kinds of observations were being performed in all three countries, with the same phenomena often observed simultaneously and independently. Thus modern astronomers have a unique and invaluable triple-check system for determining the dates and accuracy of historical records.

While the astronomical record, including the transfer of technology across national boundaries, is clear-cut for China, Korea, and Japan, less certain are the astronomical traditions in other parts of the Orient. This is especially true of Southeast Asia, where the Chinese influence is overlaid with Babylonian and other Western influences filtered through the double prism of Mohammedanism and Hinduism. Little archaeoastronomical research has been done so far in this area, but the limited results seem to support the notion that a sense of the sky is pervasive throughout all cultures.

Before war and civil strife effectively closed Cambodia to outsiders, visitors lucky enough to visit Angkor Wat were invariably struck by the extent and intricacy of this massive ruin. Built during the reign of Suryavarman II (A.D. 1113–1150) as a tribute to the Hindu god Vishnu, the vast complex of temples, galleries, walls, and courtyards covers nearly two million square meters and is laid out with such an elegant symmetry and precision that the hand of an architectural genius is obvious. Many visitors have also wondered if that architect might not have included astronomical considerations in his building plan.

Working primarily from charts and maps, since they were prevented from on-site surveys, Robert Stencel, Fred Gifford, and Eleanor Morón have concluded that Angkor Wat has "calendrical, historical, and mythological data coded into its measurements [as well as] built-in positions for lunar and solar observations." Writing in *Science*, the three researchers claim that "the sun was itself so important to the builders of the temple that even the content and position of its extensive bas reliefs are regulated by solar movement."

Angkor Wat is located in the heart of the former Khmer empire in northeast Cambodia. Basically it is constructed as a series of five concentric rectangles, including a large open courtyard surrounded by an outer moat and a series of inner temples, walls, and galleries. The outer walls extend 1,500 meters from west to east and 1,300 meters from north to south. The opposing walls of the inner galleries and temples—again oriented on the east–west axis—are so precisely constructed that their lengths vary no more than 0.1 percent.

Although the north–south axis shows no deviation from true north, the east–west axis is skewed 0.75 degrees south of east. Interestingly, at the latitude of Angkor Wat the celestial equator is shifted south by about 1.4 degrees. The slight canting of the axis thus permits an observer standing at the western end—in the entrance gate—to see the sun rise on the vernal equinox just over the central tower of the temple complex. Three days later, the sunrise can be seen over the same tower by an observer standing in the causeway a few meters in front of this entrance gate.

Stencel, Gifford, and Morón also found alignments with the winter and summer solstice sunrises, both from the same observing positions at the west gate. At the summer solstice, the sun appeared to rise directly behind a hill—the most prominent feature on the Plain of Angkor—some 17 kilometers to the northeast; and the winter solstice sun could be seen to rise behind a small outlying temple some 5.5 kilometers to the southeast of Angkor Wat.

The alignments, if they are correct, could resolve one persistent mystery about Angkor Wat. Archaeologists have long wondered why this particular temple complex faced west, when none of the hundreds of others nearby had such an orientation. The use of Angkor Wat as an observatory, or at least the incorporation of astronomical sighting devices into its architecture, could explain the unique orientation.

Unfortunately, when Stencel, Gifford, and Morón speak of the "numerical" data encapsulated in the architecture, they are on somewhat shakier ground. They propose that Angkor Wat was laid out using the ancient Khmer measurement of "hat" (approximately twice as long as the meter), so the dimensions of the buildings, walls, and galleries would represent certain basic tenets of astronomy and mythology.

For example, they claim the "total dimensions of the exterior axes of the central tower equals 90.83 hat, which is close to the average number of days (91.31) between solstices and equinoxes." Or, as another example, "the 180 hat total of the interior axes of the four axial entrances on the upper elevation symbolize one-half of the divine year of the gods (360 earth years), as well as the number of degrees traversed by celestial bodies across the sky."

Alas, the "Khmer hat" sounds remarkably similar to the "pyramid inch." Considering the nearly endless combination of walls and floors and passages in the Angkor Wat, it would be more amazing if no numerical correlations could be found.

The three researchers sound more plausible when they use astronomical orientation to explain the famous bas-reliefs of Angkor Wat. These wall carvings are truly extraordinary. They wind around the four walls of the main temple for some five hundred meters and thus represent the longest continuous set of reliefs in the world. But they have always posed a problem to archaeologists. How did one read them: clockwise or counterclockwise? And where did one begin reading?

"If read in relation to the annual and diurnal movements of the sun," say the researchers, "it seems certain that their direction is counterclockwise."

According to their premise, one would start reading on the east wall, the direction of sunrise and the spring equinox; move to the north wall, representing the summer solstice; then to the west wall, the direction of sunset and the autumnal equinox; and finally turn to the south, representing the winter solstice. In fact the Hindu conception of the solar motion was counterclockwise and the subject matter of the bas-reliefs on the four walls seems to fit this pattern.

The east-wall relief, facing the rising sun, shows "the churning sea of milk," a symbol of fecundity and new life. The west wall, or sunset side, depicts the most destructive and bloody battle of Hindu mythology. The north side, which pointed to the celestial pole, shows all the gods together, since the gods supposedly resided in this region of undying stars. Of course, for six months each year—the period coinciding with Cambodia's dry season and time of want—this north side was in shadow throughout the day. At the same time, the south side of the temple was in sunlight, and the relief there depicts the kingdom of Yama, god of death, who then would reign.

"It is not surprising that Angkor Wat integrates astronomy, the calendar, and religion," say Stencel, Gifford, and Morón, "since the priest-architects who constructed the temple conceived of all three as a unity. To the ancient Khmers, astronomy was known as the sacred science."

It is easy enough to find astronomical clues (perhaps too easy at times) in carved megaliths, stone rings, and mathematically precise temple complexes. But where does one look for clues to pretechnical astronomy among those peoples who have left no testimonies in stone? In the South Pacific, the Polynesians and Micronesians developed highly advanced skills of navigation, presumably based on positional astronomy, but they have left no visible records—no standing stones or ancient observatories—to explain the origins of their expertise, only oral traditions of legends and learning passed down from generation to generation.* Sadly, those traditions too are rapidly disappearing, victims of magnetic compasses, ship-to-shore radios, rangefinder sonar systems, and a host of other modern devices that have made an awareness of the natural world almost obsolete.

Two hundred years ago, when Captain James Cook sailed into Tahiti on his attempted circumnavigation of the globe, one of the island's navigator-priests drew up a map of his known world. The map, copies of which survive today, was very crude and had many English mistranslations of local names. Yet for all its rough nature, the map showed every major island group in the South Pacific, except Hawaii, New Zealand, and Easter Island. The Tahitian world covered a span of some 2,600 miles, about the distance between

---

* Actually, there is one megalith with possible astronomical connections in Oceania. In Tonga, a huge rough-hewn coral-rock trilithon called the *Ha'amonga a Maui* is aligned to the sunrise on the summer solstice. Some observers also think a series of very indistinct grooves in the coral may be associated with the winter solstice. The stone supposedly was erected in A.D. 1200 during the reign of the eleventh Tu'i tonga, or sacred ruler, but this date cannot be confirmed. However, if the stone appeared that late in history, then the concept of a "solstice marker" could easily have been introduced from China or Japan. Incidentally, the stone was probably quarried from coral on the sea floor in a shallow part of the lagoon. Divers incised and undercut the stone; then at low tide they lashed the rock slab to a large double-hulled canoe. When the tide rose, it lifted both boat and slab. The grooves now seen on the stone may simply be quarrying marks.

221

New York and Copenhagen—and far beyond the cruising range of any single sea-going canoe.

Western visitors from Cook's time to the present have been amazed by the maritime feats of the Pacific islanders, not the least of which is how they originally spread throughout this vast ocean world. Their forefathers probably came from the coastal island chains of Southeast Asia sometime after 1500 B.C. From almost that time until the present, Polynesian and Micronesian sailors have gone to sea in great wooden craft more than sixty feet long, with reed-mat sails, and V-shaped hulls constructed of wide planks lashed together and caulked with breadfruit sap. For stability, the Pacific seacraft were either twin-hulled or fitted with outriggers. Most Oceanic peoples adopted the double-ended Indonesian canoe design, which enabled them to change course by simply shifting the sail and the steering paddle from one end of the boat to the other. For long journeys, the Micronesians and Polynesians stocked their galleys with dried bananas, taro, fish, and fermenting breadfruit.

Using these crude craft, without any instruments, dependent only on their own senses, these people traveled freely over an area larger than the United States, Canada, and Mexico combined—an area where most landfalls were low-lying atolls virtually invisible from more than ten miles away. As David Lewis has noted in *The Voyaging Stars:* "The settlement by the Polynesians and Microne- sians of a realm that is 995 parts water for every 5 parts land is un- doubtedly one of mankind's greatest maritime accomplishments." Of course, it is possible, as he points out, to cover much of the South Pacific by short hops between islands. Using a somewhat circuitous route, one could go from Indonesia through all Mic- ronesia and Polynesia (excluding Hawaii and Easter Island) without ever making an open sea crossing of more than 310 miles. In fact, most gaps between islands, and even archipelagos, are much shorter—usually one hundred to two hundred miles at maximum.

Still, the seas are cruel, the winds are fickle, and the weather unpredictable. Mariners can easily be blown hundreds of miles off course into totally unfamiliar waters. Because getting home again is the most important part of ocean exploration, some sort of universal guidance system was needed. For many years it was assumed that the Pacific peoples based their navigation system solely on the stars. In his study of present-day, but traditional, South Sea boat pilots, Lewis found ample evidence to support these theories of Oceanic astronomy. However, he also found considerable evidence that the stellar guideposts were really only one part of an integrated and complex system utilizing a variety of natural signals and symbols.

Throughout the Pacific, the heavens were represented as a dome, or as a series of domes overlaid one within the other. Specific stars and star groups were given names, and the positions and motions of major asterisms were relatively well known to most people. However, young apprentice navigators learned a more detailed and formal version of the sky at special "schools" run by master pilots who mixed "theory" with practical experience. In the Gilbert Islands, for example, the arrangements of the rafters in the "training buildings" often represented the stars, constellations, and divisions of the sky.

The Polynesians apparently discovered early that the observed height of the pole star above the northern horizon equaled one's latitude. In other words, when Polaris was 10 degrees above the horizon at zenith, the observer was at 10° north latitude. (The general method used by Chinese and Arab sailors to measure degrees was in fingerwidths. In the Pacific, too, navigators usually judged the elevation of a star by holding their hand at arm's length with the fingers loosely extended; and this span equaled approximately 10 degrees.) While the pole-star method for determining latitude worked well when north of the equator, those people living—or sailing—south of the equator used a "zenith star" system. In brief, they learned by memory the positions of a variety of stars. When a certain star passed overhead, they could then determine their position from it. "An observer who notes that a particular star passes directly above his head will know his latitude to be the same as the celestial latitude of that star," writes Lewis.

It is unlikely that the Polynesians actually thought in terms of true longitude and latitude, however. Rather they learned, according to Lewis, "which stars are suspended over the various islands."

Thus, if a pilot saw Hokule'a (Arcturus) in the zenith, he would know that he was at the latitude of Hawaii. Similarly, if Sirius (declination 17° south) passed directly overhead, he knew he was at the same latitude as Tahiti and Fiji.

Although the zenith stars provided a rough idea of a navigator's position, the actual steering was probably done by horizon stars. One informant told Lewis that "a *fanakenga* star is one that points down to an island, its overhead star. When it is nearly overhead, it indicates that you are reaching the island. It is quite different, of course, from the *kaveinga*, or compass star, low down on the horizon, that you steer by."

Most of the island navigators maintained some sort of "star compasses." In the Caroline Islands, these compasses became quite complex and detailed, consisting of as many as thirty-two points, marking both the rising and setting positions of the star on a stylized representation of the horizon circle. According to Lewis, the star compasses were simply systematized versions of the informal paths traced across the sky by the *fanakenga* stars. The rising and setting points indicated general directions on the horizon rather than specific islands. This technique of navigation seemed to be used throughout all the island groups, although the "compasses" were not always formally drawn out.

But the seafarers did not steer by the stars alone. How could they under adverse weather conditions or during daylight? The Polynesians, for one, developed a wind compass that marked the directions of the prevailing winds. In fact, almost all navigation techniques emphasized the interpretation and analysis of currents and wave patterns. As William Alkire has noted in his *Introduction to the Peoples and Culture of Micronesia:*

> In the Marshall Islands, traditional voyaging was generally north and south. . . . The archipelago lies at right angles to the prevailing winds and the easterly swell carried before the winds. Since the atolls are generally quite long and closely spaced, they interrupt the smooth transit of the swell; therefore, the basis of the Marshallese navigation focused on interpreting the patterns, orientation, intensity, and direction of

deflected waves. The easterly swell is deflected in a predictable and constant pattern as it passes through the islands; the trained navigator therefore is able to judge his position when out of sight of land by reading surrounding wave patterns and conceptually tracing the origins of the particular pattern back to a known reef or island of the chain.

The famous Marshall "stick charts," now sold throughout the South Pacific to tourists as "Polynesian maps of the sea," are neither charts nor maps. Rather they were diagrams of wave patterns used primarily as teaching devices for apprentice navigators. According to Alkire,

A chart was made from a number of interconnected thin strips of wood ordered and affixed to each other to represent the pattern of waves between and around several islands and reefs (represented on the chart by small shells or coral pebbles tied to the strips). The navigator identified the islands on the chart, then set a course from one to the other, observing the pattern of curved and intersecting strips representing waves and currents he would encounter on the particular voyage.

Even with their wind, wave, and star compasses, the South Pacific sailors may not have been all that precise in steering toward a destination. But they didn't really need to be. In Polynesian navigation, a close encounter was just as good as a direct hit. One informant told David Lewis: "You see those *puko* trees? We have a proverb which says, 'It is enough that we strike the row of *puko* trees. You need not hit a particular tree.' In the same way, a canoe captain would aim for the middle of a group, rather than for a particular island."

This technique is known to sailors as "expanding the target," and Lewis finds it widespread throughout both Polynesia and Micronesia, with only the terminology changing from area to area. In the Carolines, for example, the "trees" become "bird zones" or "submerged reefs." In short, the South Pacific islanders used a multitude of signals and signs to help them sail from island to island with apparent ease: They read wave patterns and wind directions, they followed the flights of land birds, they observed clouds build-ing over distant specks of land, they followed the sun during the day, and they felt the humidity of winds blowing over their shoul-ders. And, of course, they watched the stars.

But did they ever develop a science of astronomy? No, not in the strict academic sense. As one of Captain Cook's scientific voyagers noted two hundred years ago, the astronomy of Polynesia was completely subservient to navigation.

Ironically, by putting their observations of the stars to a purely practical use, these supposedly simple, naturalistic people may have been closer in philosophy to modern man than the sophis-ticated and cultured Chinese court astronomers whose excellent positional observations were totally subservient to astrology.

The Past
Is Future

"The maturation of archaeoastronomy as an accepted branch of science may be regarded as one attempt by science to seek its earliest roots—the roots of a cultural tradition that knew no ethnic or geographic boundaries and, that, in time, transformed the nature of human society."
—Boyce Rensberger

"Of course, it has changed my life. But I suspect it changed archaeoastronomy as well. Otherwise, you wouldn't be writing this book." Sitting in his Washington office, Gerald Hawkins is talking about the chain of events following his pioneering research at Stonehenge. Nearly two decades had passed since I wrote my news release about his analysis of the astronomical alignments at Stonehenge, and now I was back where I—and this new discipline—had begun.

In the months and years following publication of his article in *Nature*, Hawkins was deluged with calls and letters from supporters and detractors around the world. His book on the subject turned into a best-seller. A television special on Stonehenge was, at the time, the most widely viewed documentary ever broadcast. He produced a second book to answer critics while encouraging further discussion and, as its title implied, to extend his own research "beyond Stonehenge." And scores of standard reference books have been rewritten to reflect the new understanding of that ancient site.

Hawkins also started a long and far-reaching debate between astronomers and archaeologists, a debate that would eventually bring about a synthesis of several once-disparate points of view. Yet today, an academic generation later, he is still recognized—and identified—as "the man who used the computer to show Stonehenge was an observatory."

Today, too, Hawkins has exchanged his government service with the Smithsonian Astrophysical Observatory for a position with an agency of the U.S. State Department. He continues to lecture in astronomy programs at the Smithsonian Institution and is affiliated with the Center for Archaeoastronomy at the University of Maryland, where he served as an advisor in setting up the complicated equations for a computer program used to check the astronomical alignments of suspected ancient observatories. (The Center hopes this "streamlining" of the mathematics may assist the research of both the astronomical specialist and the usually less mathematically oriented anthropologists and archaeologists.)

In addition to some continuing field research on the stone rings of Britain, he has also spent considerable time giving advice and counsel to other researchers in the rapidly growing discipline of archaeoastronomy. News reporters still call him regularly on June 21 with questions about the significance of stone alignments and the summer solstice. As one speaker at a 1979 conference noted: "I would call Hawkins the Father of Archaeoastronomy, except that he looks too young."

The recognition of Hawkins's role as founder is appropriate; for, if archaeoastronomy has changed his life, he almost single-handedly changed public and scientific conceptions of archaeoastronomy. He brought it out of the academic closet.

From its one-time status as a fringe science, fraught with fallacies and plagued by phonies, archaeoastronomy has become almost overnight a full-fledged, officially accepted scientific endeavor, with its own specialized journal and national conventions complete with academic debates over protocol, procedures, and the "institutionalization of methodology," a sure sign that the discipline has entered the mainstream. Articles by archaeoastronomers, while still appearing regularly in popular magazines for a curious public, are also accepted readily by the traditional, refereed, scientific journals. Archaeoastronomy has become the hottest growth stock in science.

The marriage between astronomy and archaeology has not been an easy one, however. The two parties still are not completely compatible. As Jonathan Reyman has observed in his review of archaeoastronomy published in the *Archaeoastronomy Bulletin:*

> Astronomers are often ill-prepared to understand the archaeology of a site, while archaeologists' ignorance of astronomy often leads them to make erroneous statements about astronomical phenomena. There is also the problem that modern astronomy in the western world has little to do with day-to-day subsistence behavior, whereas for the most part non-industrial peoples' astronomy is an integral part of the subsistence strategy. . . . Astronomers often tend to ignore this aspect of culture. . . . Archaeologists, on the other hand, often do not understand what is and what is not possible in terms of observational astronomy and record-keeping. Therefore they may ascribe to peoples astronomical achievements which are in fact beyond the group's technical capabilities.

Astronomers also find the ambiguous, uncertain, and usually unmeasurable aspects of archaeology disturbing. Long used to working with physical constants and known properties that can be expressed mathematically, they try to apply the same standards to archaeological sites, sometimes with disappointing results. Most sites defy quantitative analysis. Excavation records are fragmentary, site maps inaccurate, construction dates approximate. Moreover, many sites have been altered considerably over time by peoples who followed the original observatory builders, adding features or structures unrelated to the builders' astronomical markers. The surrounding terrain may have changed, too, with natural horizon-sighting points removed or altered.

The establishment of criteria for field research has understandably been one of the most pressing needs for archaeoastronomers. As early as 1966, in his Smithsonian report, *Astro-Archaeology*, Hawkins prepared a set of five guidelines to help normally skyward-gazing astronomers look more carefully at the ground. He revised the list slightly in 1979, and added a sixth point.

1. The construction date of a site should not be determined from the supposed astronomical alignments. The date of a site must be established first by other means, then the researcher can check for any alignments with star positions of that epoch. (As mentioned earlier, Lockyer and other early archaeoastronomers followed the opposite track; for example, if the stones pointed to Aldebaran's position in 3000 B.C., then they assumed that the stones were erected then.)

2. All alignments at a site should be restricted to manmade markers. Says Hawkins: "It may be true that for a given location the moon at a particular point in its cycle rises over a prominent hill on the horizon, but this is not sufficient evidence that the builders were aiming at the moonrise. On the other hand, a distant hill marked by a row of stones can be valid, since the man-made alignment establishes the first-order significance."

3. Alignments should be postulated only for a homogeneous group of markers—that is, stones should not be mixed with grave mounds, or round barrows with tall posts. Otherwise, Hawkins warns, "the number of alignments will increase as the number of possible markers increases until it becomes meaningless." Also, the markers should all be related to each other architecturally or culturally.

4. All related celestial positions should be included in the analysis. If the midsummer sunrise is found as one alignment, then the three other solstice points (summer sunset, midwinter sunrise and sunset) should also be sought. If a single bright star is proposed as a target, then all stars brighter, or of comparable brightness, should be measured too.

5. All possible alignments at a site must be considered and tested.

6. All alignments should be as accurate as an investigator can make them. If any measurements cannot be made with a high degree of precision, then the researcher should reduce the reliability of that aspect of the site.

Ideally, all the criteria should be met. Because this is hardly ever possible, the conclusions from any given site become progressively less reliable when one or more are not fully met. At the urging of the archaeoastronomy community, Hawkins assigned a proportional weight to each of the criteria, with a maximum "reliability value" of 100.

Astro-date fits archaeology:	20
Markers are anthropic:	20
Markers are homogeneous:	20
Astro-pattern is homogeneous:	20
All alignments are considered:	10
Alignments are accurate:	10
	100

On this scale, Hawkins feels his own Stonehenge research would score about 80; but the work of Alexander Thom on some stone rings—for example, the one at Kintraw—would score no more than 30. According to a personal note from Hawkins, "Stonehenge does not quite score 100 because the alignments are not placed as accurately as the builders could have done. Why they settled for an accuracy of a degree or so we cannot tell. Thom's work scores lower because the notches on the horizon are not anthropic and are not homogeneous with the rows of stones."

In less than two decades, the study of the connection between stones and stars has evolved from a shadowy, slightly disreputable, and sometimes daffy pursuit of particular appeal to what John Mitchell has described as "elderly scholars of sage disposition" into a dynamic field concerned with accuracy, consistency, and high scientific standards. But what does it all mean, this accuracy in alignments, this careful dating of stellar epochs, this concern that all markers be man-made? And more to the point, why do we care if the Pawnees observed the stars over the Great Plains, or the Mayas charted Venus, or the priests of Angkor Wat greeted the sun on the midsummer's morn?

Archaeoastronomy not only adds to the general chronicles of human intellectual history, but it also makes more specific contributions to the study of our past—and future—both in the sky and on the earth. Modern astronomers have found that the observations made by ancient peoples—no matter why they made them—may contain valuable clues to questions of current astrophysical interest. Since the record of certain phenomena, such as supernovae, can be extended several thousand years farther back in time, one can check to see if the frequency of star explosions has changed significantly. Also, by identifying the observations of supernovae and comparing the dates of the original explosions with the remnants of those stars seen today, it may be possible to determine how fast this star debris is being propelled through space. Not only does this tell astrophysicists how much energy is needed to blow a star apart, but the expansion rate of the dust and gas can give some clues to the density and composition of the near-vacuum space between the stars.

The answer to another perplexing puzzle involving a star may also come from archaeoastronomy, although the research so far has only deepened the mystery. From 1000 B.C. to A.D. 200, almost every literary reference to the bright star Sirius has mentioned its reddish color. However, if you look now at this star, the brightest in the sky, it is obviously blue-white, not red. Why isn't the star the same color now as it was in antiquity?

Kenneth Brecher has carefully sought out the historical references to the color of Sirius. The earliest is found on a Babylonian cuneiform tablet dating from about 700 B.C., which describes it as "rising in late autumn and shining like copper." It is strange that the Egyptians, who faithfully watched the heliacal rising of Sirius as the signal for annual Nile floods, never recorded its color. The Romans, however, clearly saw red when they looked at Sirius, calling it *Rubra Canicula*, or the "red dog." The clearest and most unambiguous reference is from Ptolemy, the Greek astronomer who shaped so much of medieval science. He listed the six brightest red stars in the heavens as Aldebaran, Betelgeuse, Antares, Arcturus, Pollux, and Sirius. Today, the first five stars are still seen as red to the naked eye, and astronomers classify them as "red giants." Sirius, however, is definitely white, a fairly average star much like our own sun.

Could Ptolemy and the observers of the past have been wrong? Did they simply get their colors mixed up? Brecher thinks this is unlikely, considering the brightness of Sirius, its importance to so many cultures, and the general accuracy in reporting the colors of other stars of lesser magnitude. Like the sun and moon, stars often appear reddish when low on the horizon, since the earth's atmosphere filters out the blue portion of their spectrum. Because the heliacal rising of Sirius was, particularly for the Egyptians, among the most important celestial events, the early reporters may have been referring to its color at sunrise. This sounds plausible, says Brecher, except that other stars seen at the horizon were usually described by their true colors. The written historical record, then, suggests that Sirius was red as recently as two thousand years ago, a fact that seems to defy every known law of stellar physics.

In his excellent study in *Astronomy of the Ancients*, Brecher also offered several possible astronomical explanations for the star's apparent change in color: Sirius may have been covered by an interstellar-dust cloud that caused it to look red. Or maybe Sirius B, the tiny white-dwarf star in the Sirius binary system and the invisible companion cited by the Dogon, was a red giant two thousand years ago. Or maybe vast amounts of mass were somehow transferred from Sirius A to Sirius B, causing the white dwarf to become temporarily a red giant of sorts, or what Brecher calls a slow nova.

If Sirius B had been a red giant in orbit around Sirius A, the two stars might have appeared from earth as a single point of light that changed color within a period of about fifty years. Observations of this changing color could have become mixed with legends about dying stars and then spread across Africa from the Egyptians to the Dogon, who, in turn, could have changed the "red star" into an unseen "dark companion" as Sirius B gradually faded from view.

233

Brecher is the first to admit that his theories do not satisfy modern astrophysicists. Indeed, each theory requires stretching the laws of physics almost to the breaking point. There still is no known process by which a star (Sirius B) could evolve from red giant to white dwarf in only two thousand years.

"The redness of Sirius creates an incredible problem for theoretical astrophysicists, whose understanding of stellar evolution comes as much from computer calculations as from observations of the sky," says Brecher. "It's my secret hope, nevertheless, that Sirius truly *was* a red star in historical times. I would much prefer to learn stellar evolution from the ancient myths of man than from the modern myths of the computer."

Sirius may not be the only star that has changed color in a short time. Brecher has also found that Betelgeuse, now seen as red, was apparently observed as a white or yellow star by Chinese court astronomers around 100 B.C. This information comes from an ancient document, the *Shih Chi*, and was pointed out to Brecher by two Chinese colleagues. Oddly enough, within only 250 years of that Chinese observation Ptolemy cited Betelgeuse as one of the most prominent *red* stars in the sky. Obviously, it is impossible for a star to evolve to a new state—or change its color—in such a short period. However, there is a huge cloud of gas and dust now seen moving away from Betelgeuse, and modern measurements indicate this gaseous shell was probably ejected from the star some twenty-seven hundred years ago, or only a few centuries before the Chinese observations. Brecher suggests the ejection might have temporarily heated the star's photosphere, thus making it appear white to the Chinese, while rapid cooling would have made it appear red to Ptolemy in A.D. 150.

Astronomers also are searching the ancient sky records to find out something about the earth, for references to solar eclipses in the past may indicate how much the earth's rate of rotation is slowing down today. Initially, the interest in solar eclipses was purely historical: some major events of antiquity could be more precisely dated by correlating them with known solar eclipses traced back through time by means of the saros cycles. While analyzing the

234

eclipse records, astronomers discovered many eclipses were reported thousands of miles away from the location predicted by modern calculations. Obviously, something was wrong.

Subsequent experiments have shown the earth's spin is decreasing over time, so that the length of the day is increasing by a few thousandths of a second each century. A variety of theories have been suggested to explain the slowdown: solar tides, ocean tidal action, a shift of mass in the earth's core, or even a general weakening of the force of gravity throughout the universe. If this slowing of the earth's rotation was enough, an eclipse expected to occur over Damascus might have been seen in New Delhi instead. By measuring the displacement between an observed eclipse path and the predicted path (as calculated for an earth spinning at a constant rate) astronomers may be able to determine the exact rate by which the earth has slowed since that time in history. Calibration of the current precisely timed rate of spin with the past rate may allow scientists to predict the continued spin-down for the future—and determine its implications for life on this planet.

A concerted search has been made through records of the late Babylonian period (450–50 B.C.), through Greek and Roman classics, Chinese annals from 100 B.C. to A.D. 1000, and the medieval European town and monastic chronicles of A.D. 500–1200. (So far, the Mayan records have proven less than helpful because of problems in correlating their calendar with the dates of known eclipses over the Yucatan.) To be usable, the ancient records must provide a location for which a longitude and latitude can be assigned. (The band of totality is usually no more than one hundred miles wide, so a difference of one hundred miles or more in the path of an observed eclipse could mean two separate eclipses several centuries apart.) They should also give a clear indication that the eclipse was total. (Partial eclipses can be seen several hundred miles on either side of the total eclipse path, so observations of a partial eclipse could refer to several different events.) The record should show a date that can be compared with modern calendars and a north–south path, for this is the type of path that would show the most displacement by a slowdown in the earth's west-to-east motion.

Ancient eclipse records may also help scientists tell us something about the future of the sun itself. John Eddy, an expert on Amerindian astronomy, is also a physicist who has been searching for clues to solar behavior in ancient records. It was Eddy who re-identified the "Maunder Minimum," that period of unusually low sunspot activity during the late seventeenth and the eighteenth centuries, and showed it corresponded with a period of extremely cold weather on earth, a time sometimes called the Little Ice Age.

Eddy maintains that the sun's activity shows long-term variability beyond the well-known eleven-year cycle, with periods ranging from several hundred years to perhaps as long as a thousand years. The effect on earth's climate of prolonged low solar energy—if the Maunder Minimum is any indication—could be significant and possibly catastrophic.

Through an analysis of more modern observatory records from both Britain and the United States, Eddy made another startling discovery: The sun's diameter seems to be shrinking at a rate of about eight miles per year. Considering the sun's massive size, this shrinkage is small, but in terms of energy output it is substantial. Again, it could have serious implications for the climate—and ultimate fate—of earth.

To check his theory and to make sure the shrinking was not simply an apparent effect due to either changes in the earth's atmosphere or telescopic effects, Eddy looked at records of eclipse observations from medieval Europe. He found Clavius, an Italian astronomer, had observed and described in detail an eclipse seen in 1567 over Rome as "annular"—that is, a thin ring of sunlight had remained around the moon's disk. Oddly enough, modern calculations based on the size of the sun today indicate this eclipse should have been total (as predicted). Is it possible that the sun was slightly larger then, so that its apparent disk was not covered by the moon? Eddy thinks this is a distinct possibility, but additional examples of other unexpected and unpredicted annular eclipses must be sought in historical records.

Rolf Sinclair of the National Science Foundation also has suggested that clues to the cycles of solar variation might be found in ancient art depicting eclipses.

236

Without doubt, solar eclipses are among nature's most spectacular sights. As the moon's disk slowly slips in front of the sun, the bright solar disk is reduced to a crescent, the sky darkens, and the temperature drops. As totality nears, birds roost, livestock return home from pasture, and flowers close their petals. Just before the last thin crescent of sun disappears, a string of bright points, known as Baily's beads, appears around the moon's disk. This phenomenon is caused by the slanting rays of the sun shining through the valleys and mountains on the edge of the moon. There is a final, brilliant flash of light—the breathtaking diamond-ring effect—and then the world is plunged into darkness. At this very moment, an eerie, shimmering, pearl-gray halo suddenly flares out from around the darkened sun. This is the corona, or the sun's tenuous outer atmosphere of superheated gases, which cannot be seen at any other time. This glowing, pulsating halo, sometimes shot through with thin red streaks caused by explosions on the sun's surface, varies in brightness and structure depending on the amount of solar activity at the time of the eclipse (see Figure 69).

The corona is actually best seen with the naked eye, for this human organ is much more sensitive to faint shadings and delicate motions than most films or electronic detectors. Sinclair therefore thinks it is reasonable to expect that ancient artists recorded the variations in coronal brightness and structure seen during different eclipses.

The sun disk with wings and tail plumes appearing in both Egyptian and Babylonian art could be the representation of a total eclipse and corona seen at sunspot minimum. (Similar designs survive in modern times; for example, the sunburst symbol of the Smithsonian Institution and the popular ceramic suns from Metepec, Mexico, both show jagged-edged ray patterns that may have originated as stylized versions of a coronal halo—see Figure 70).

Sinclair suggests correlating known observations of eclipses back to 1200 B.C. with changing styles in artistic representation to produce a rough record of the sun's changing activity. As he notes, it could "give information on solar activity (or at least set bounds on it) up to the time when the solar symbols became conventionalized." Moreover, as Sinclair wrote in *Science,*

There is a distinct possibility that the challenge of coming to grips with total eclipses—which occurred so rarely as to be of tremendous and unexpected impact, but yet just often enough to stay in legend and history and have their effect reinforced—gave rise to concepts of religion and science that altered the course of civilization irrevocably and raised it to new

Figure 69. *A total solar eclipse showing the solar corona and distinct features of activity on the sun's surface, including flares and prominences. Rolf Sinclair has suggested that a long-term history of the sun's variability might be determined from a study of how ancient artists depicted the coronal appearance at totality.* Photo courtesy Harvard College Observatory

levels. We already know that recording the frustrating details of the moon's motion became the organizing theme for the people of the British Isles in the second and first millennium B.C. And the solution of the apparent motions of the planets—seemingly so regular, and yet irregular enough to be a just-solvable problem—triggered modern science.

Indeed, the greatest contribution of archaeoastronomy may be that it provides an overall vision of man's continuing relationship with nature. As Jonathan Reyman writes in the *Archaeoastronomy Bulletin:*

> The discipline is at a crucial point. The existence of astronomical record-keeping and astronomically aligned architectural features have been adequately demonstrated. The problems remaining concern the significance of these records, and the demonstration that the findings are important for the study of prehistoric behavior.

Figure 70. *These two modern and stylized representations of the sun may stem from the ancient tradition of depicting the coronal flares seen during total solar eclipses.* Left: *A ceramic folk art sun face from Metepec, Mexico;* right: *The symbol of the Smithsonian Institution.*

Many scientists still feel the evidence of an astronomical awareness among preliterate peoples is interesting only as a minor phenomenon. But this attitude is changing as more and more evidence
reveals the central role astronomy played in almost every early civilization—and especially in marginal agricultural societies.

For ancient man, astronomy was first a technique of survival. By
marking the sun's swing south and north on his horizon, the primitive farmer could tell the proper times for preparing fields, planting seeds, and harvesting crops. His life became inexorably tied to
the solar cycle; it shaped his religion, his politics, and his societal
structure in ways we are only now realizing. Archaeoastronomy is
one means of understanding early peoples on these terms. For example, realization that the Plains Indians were precise observers of
the skies changes our conception of them and explains many
aspects of their lives, such as how towns and ceremonial centers
were laid out in reference to the summer solstice. The evidence of
sun-watching at Casa Grande can now be seen as related to Mesoamerican traditions, thus suggesting that contacts, including trade
and commerce, between the American Southwest and the Valley
of Mexico were more extensive than previously thought. Perhaps
there was even a reverse migration from Mexico north into the
Arizona desert that can be traced with astronomical clues. As more
sites are found, explored, and dated they may help clarify the way
prehistoric man dispersed throughout North America. The sites
provide, as Eddy has called it, "a trail of crumbs that we can follow
like Hansel and Gretel back to their ideological beginnings."

In the broadest sense, archaeoastronomy also has pushed back
all the clocks of human history. Marshack's work on Neolithic artifacts suggests that cognitive ability emerged very early in human
history and that the recording of celestial events—particularly
those occurring in cycles—may have provided the foundations for
all science by establishing concepts of time-factoring, counting,
and recording. And Hudson's research on sky-watching by the Indians of pre-Conquest California supports this premise by showing
that the "roots of science, such as calendrics and astronomy, are
not invariably the intellectual achievements of food producers."

240

On the contrary, a rough grasp of the relationship between the changing positions of certain celestial objects and the changing seasons may be a necessary first step before hunter-gatherers become agriculturists. In short, almost every discovery in archaeoastronomy, from the medicine wheels of the American Plains to the stone calendar of Kenya, adds a new dimension to human history.

Astronomical concepts previously considered the exclusive product of modern science also have been found to exist among preliterate cultures, hinting at if not direct knowledge, remarkable intuitive powers. For example, Anthony Aveni and Thomas Zuidema have found evidence that the Incas marked the antizenith days, those dates when the sun at midnight passed directly below an observer at Cuzco. The Incas apparently realized that the phases of the moon were somehow related to the nighttime direction of the sun. They may not have known that the moon was illuminated by the sun, but they sensed that the two bodies were linked. A key event, then, was the midnight passage of the full moon through the zenith, for it coincided with the sun's antizenith point. (Gary Urton has found this up-and-down cosmology persisting among several South American Indian tribes. The Kogi of Colombia view the universe as two cones joined at their bases on earth. One cone points into the heavens and represents day; the other points down into the earth and represents night.)

Although there is no clear evidence that they ever did so, Aveni feels the Incas certainly had a unique opportunity to determine the rotundity of the earth from some very simple observations. This stems from the unusual geographical orientation of the Inca empire. The main axis, political as well as geographical, ran north and south; and, at the height of its power, the empire extended over nearly 30 degrees of latitude. Theoretically, an Inca astronomer who traveled regularly between Quito in the north and the Chilean outposts in the south would have noted a marked change in the stars, both in the position of familiar constellations overhead and the appearance of new stars along the northern and southern horizons. Could the astronomer also have deduced something about the shape of the earth from these observations? Perhaps, says Aveni; but any record of such a landmark discovery is missing. All that exists today are the tantalizing hints offered by the Inca's ceremonial celebration of the antizenith points.

Geography may have played an even larger role in shaping the ways that all ancient societies observed the sky. For example, those peoples living in the temperate zones invariably looked to the horizon for their celestial guideposts, while those peoples living in the tropics looked directly overhead. Kenneth Brecher and Philip Morrison were among the first modern researchers to suggest that ancient astronomy, and especially solar observations, might be divided into two classes: horizon-based and zenith-based (although apparently Zelia Nuttall, a Harvard anthropologist, made a similar suggestion nearly a century ago). In a 1978 paper published in the *Bulletin of the American Astronomical Society*, they proposed that

> . . . cultures living in open country at high geographical latitudes with a strongly varying day length and temperature, and moderately unobstructed horizons, find solstitial indicators useful. By contrast, in low geographic latitudes, there is little daylight and temperature variation to draw attention to the solstices, though the year naturally divides into rainy and dry seasons. In mountainous tropical regions with topographically complex horizons, the twice-yearly zenith transit is nonetheless easily measurable.

Well-known examples of horizon-based astronomy abound in the northern temperate zone: Stonehenge, Callanish, the medicine wheels of the Great Plains. But equal numbers of zenith-oriented sites may be found in the tropics: the zenith-passage "sighting hole" at Monte Albán, the alignment of many Mesoamerican cities to the sunrise-sunset direction on the solar transit day, the zenith-star navigation systems of the South Pacific. Brecher and Morrison also cite "the major alignment with the sunset direction on August 13, the zenith passage date, at the latitude of the early Mayan city of Copán." And they find "striking evidence for an Inca concern with the zenith transit of the sun, as revealed both by measuring devices (*intihuatanas*) and from Spanish texts" (see Figure 71).

Figure 71. *The possible astronomical use of the "hitching post of the sun" at the Inca mountaintop ruins of Machu Picchu has long been the subject of speculation. Kenneth Brecher proposes that this monument might have been used as an* intihuatana, *or measuring device, for marking the disappearance of shadows at the moment of the sun's noontime zenith transit over this location. As Brecher noted in a 1977 paper presented to the American Anthropological Association, "It is unaligned with any obvious astronomical direction, it bears no azimuthal markings, and its form allows a practical zenith passage determination."*

Brecher also points out that the "dates of the two zenith transits at the latitude of Cuzco (October 30 and February 13) agree with the onset of the rainy season and harvest, respectively, whereas the solstice and the equinox dates have no equally conspicuous significance."

Naturally, there are exceptions to this strict horizon-zenith division of ancient astronomy; the recent discovery of a noontime solar marker for the solstices at Fajada Butte, New Mexico, is one of the more outstanding examples. Yet it seems clear that the way ancient man observed the sky—and measured his time—may have been strongly influenced by his geographical location.

Still, no matter where a civilization began, its people had an overpowering urge to look up at the heavens. Indeed, the universality of astronomy among peoples separated by thousands of miles and thousands of years testifies to the tremendous grip with which the stars have always held mankind. While diffusionists may claim the common factors in so many cosmologies prove that all astronomy sprang from a common source, it is more likely that persons of intelligence in every society simply reached similar conclusions through similar observations. The basic celestial phenomena have remained unchanged for eons, and presumably an emerging civilization today might react to the heavens in much the way Cro-Magnon peoples did some thirty thousand years ago.

Similarly, the spontaneous and independent appearance of astronomy among so many different peoples suggests it is a very natural and very human response. It is also an effective rebuttal to the claims of Erich Von Däniken and his followers that interstellar space travelers brought astronomical knowledge to earthbound dolts. Marshack's research, Thom's studies of the megalithic stone circles, and Eddy's findings on medicine wheels all demonstrate that so-called Stone Age people, with limited resources, no written language, and only the crudest tools, could develop a basic astronomy—untaught by anyone and unaided by anything other than their own senses.

Archaeoastronomy may have struck another, more subtle, blow to Von Däniken. His great success and popularity with the general public have been due in part to the inability of legitimate scientists

244

to convey the excitement and romance of science. Many professionals—fearing peer disapproval, perhaps—are reluctant to write popularized versions of their work. (In fairness, too, the demands of teaching, research, and seeking additional funds also take much effort for these professionals and leave little time for popularization.)

The nature of archaeology has also changed in the twentieth century. The era of great discoveries has long passed; this is a period of refinement, categorization, and careful chronology—hardly the stuff of dreams. Rightly or not, the public still remembers the era of great discoveries in the late nineteenth and early twentieth centuries, when dashing explorer-adventurer-scientists made fabulous finds of treasure troves in exotic lands. These discoveries created a sense of adventure and excitement that the public came to expect. When modern science would not—or could not—provide this "thrill of archaeology," Von Däniken and his imitators did.

Part of archaeoastronomy's new appeal to the public may be that it restores some of the romance and excitement missing from much of modern research. Certainly the subject matter is fascinating: ancient peoples performing ritualistic science in exotic settings. Chichén-Itzá, Stonehenge, Tikal, Namoratunga conjure up exciting images. In addition to these remote lands and this esoteric activity, a very real human factor is always present in this science.

Traditional archaeology, originating with amateurs and later adopted as an adjunct to the academic study of history, is concerned primarily with *when* things happened—events as revealed by silent stones: the fall of Rome, the rise of the Assyrian empire, the battle of Thermopylae. By contrast, archaeoastronomy, particularly as practiced in the United States, springs from anthropology, and is therefore concerned more with *why* things happen—and with the people who made them happen. Archaeoastronomy seeks to explain how people lived and thought in the past by exploring what E. C. Krupp has called "the interaction between brain and sky."

The dusty, dry, and dull stratification layers of the excavation pit, the colorless chronologies of kings and court officials, the lists of battles and administrative decisions become real human dramas in archaeoastronomy. This is a science concerned with the struggle of men and women against hostile environments. It is a science that chronicles both the intellectual development of all human civilizations . . . and the psychic fulfillment of individuals. And it is a science that explores the deepest human emotions, for it seeks to understand the relationship between humans and nature. Oddly, that relationship has not changed much over thousands of years. We are closer to our ancestors than we imagined.

At some time, all thinking beings must ask why they are here. Why has this tiny speck of dirt and rock called Earth been set spinning through space? Why has this one planet been chosen as the abode for life? What is its relationship to the million stars above? What future can be discerned in the slow paths those flickering lights trace across the sky: Where am I? Who am I? Why am I?

As we stand beneath a star-dappled sky and stare into space, we are linked in mind and spirit to those other watchers of the skies, who, in millennia long gone by, gazed upon the same stars and perhaps asked the same questions.

Bibliography

Abetti, Giorgio. *The History of Astronomy*. New York: Henry Schuman, 1952.

Alkire, William H. *An Introduction to the Peoples and Cultures of Micronesia*. 2nd ed., Menlo Park, California: Cummings, 1977.

Allen, David A. "An Astronomer's Impression of Ancient Egypt." *Sky and Telescope*, July 1977, pp. 15–19.

Arochi, Luis Enrique. "The Visit of the Plumed Serpent." *Americas*, 29(8) [August 1977], pp. 33–35.

———. "A Star Is Reborn." *Scientific American*, 239(4) [October 1978], p. 102.

Atkinson, R. J. C. "Interpreting Stonehenge." *Nature*, 265 (January 6, 1977), p. 11.

Aveni, Anthony, ed. *Archaeoastronomy in Pre-Columbian America*. Austin: University of Texas, 1975.

Bibliography

Aveni, Anthony, ed. *Native American Astronomy.* Austin: University of Texas, 1977.

———. "Old and New World Naked-Eye Astronomy." *Technology Review*, November 1978, pp. 60–72.

———, Hartung, Horst; and Buckingham, Beth. "The Pecked Cross Symbol in Ancient Mesoamerica." *Science*, 202(4365) [20 October 1978], pp. 267–79.

Baity, Elizabeth Chesley. "Archaeoastronomy and Ethnoastronomy So Far." *Current Anthropology*, 14(4) [October 1973].

Beach, A. D. "Stonehenge I and Lunar Dynamics." *Nature*, 265 (6 January 1977), pp. 17–21.

Bord, Janet; and Bord, Colin. *The Secret Country.* New York: Walker, 1976.

Brandt, John C. "Pictographs and Petroglyphs of the Southwest Indians." *Technology Review*, 80(2) [December 1977], pp. 29–39.

———, and Williamson, Ray A. "The 1054 Supernova and Native American Rock Art." *Archaeoastronomy*, 10 (supplement), 1 [1979], pp. 1–38.

Brecher, Kenneth. "Sirius Enigmas." *Technology Review*, 80(2) [December 1977], pp. 52–63.

———. "Solar Zenith Markers in Ancient Astronomy," paper presented to American Anthropological Association Annual Meeting, 1977.

——— and Feirtag, Michael, eds. *Astronomy of the Ancients.* Cambridge, Mass.: MIT Press, 1979.

———, and Morrison, Philip. "The Solar Zenith Transit in Ancient Astronomy," Abstract, *Bulletin of the American Astronomical Society*, 10(1978), p. 610.

Brown, Peter Lancaster. *Megaliths, Myths, and Men.* New York: Taplinger, 1976.

———. *Megaliths and Masterminds*, New York: Charles Scribner's Sons, 1979.

Brugsch, Heinrich. "Astronomical and Astrological Inscriptions on Ancient-Egyptian Monuments." Los Angeles: *Griffith Observer*, 1978–79.

248

Bush, Sherida. "Ancient Astronomer—Fact or Fancy?" *Psychology Today*, 10(7) [December 1976], p. 100.

Callaghan, Catherine. "Comment on Turton and Ruggles." *Current Anthropology*, 19(3) [September 1978], p. 603.

Carlson, John B. "Maya City Planning and Archaeoastronomy." *Archaeoastronomy Bulletin*, (3) [May 1978], pp. 4–5.

————. "The Nature of Mesoamerican Astronomy: A Look at the Native Texts." Abstract, Proceedings of the American Association for Advancement of Science. San Francisco: January 1980.

Chamberlain, Von Del. *Pawnee Stars*. Santa Barbara: Ballena Press, in press.

Chou, Hung-hsiang. "Chinese Oracle Bones." *Scientific American*, 240(4) [April 1979], pp. 134–49.

Collea, Beth A., and Aveni, Anthony. *A Selected Bibliography on Native American Astronomy*. Hamilton, New York: Colgate University, 1978.

D'Ambrosio, Ubiratan. "Science and Technology in Latin America During Its Discovery." *The Impact of Science on Society*, 27(3) [1977], pp. 267–74.

de Camp, L. Sprague. *The Ancient Engineers*. New York: Ballantine, 1974.

de Landa, Diego. *Yucatan Before and After the Conquest*. New York: Dover, 1978.

de Santillana, Giorgio, and von Dechend, Hertha. *Hamlet's Mill*. Boston: David R. Godine, 1977.

Douglas, John. "The Origins of Culture." *Science News*, 115 (14 April 1979), pp. 252–54.

Doyel, David E. "The Prehistoric Hohokam of the Arizona Desert." *American Scientist*, 67(5) [September–October 1979], pp. 544–51.

"Earliest Recorded Eclipse." *Sky and Telescope*, 41(1) [January 1971], p. 26.

Eddy, John A. "Medicine Wheels and Plains Indian Astronomy." *Technology Review*, 80(2) [December 1977], pp. 18–31.

Egger, Herman. "More About Sundials—Old and New." *Sky and Telescope*, 42(5) [May 1972], pp. 288–89.

Bibliography

Enciso, Jorge. *Design Motifs of Ancient Mexico.* New York: Dover, 1953.

Fagan, Brian. *Elusive Treasure.* New York: Charles Scribner's Sons, 1977.

Frazier, Kendrick. "Solstice-Watchers of Chaco." *Science News,* 114(9) [26 August 1978], pp. 148–51.

Gingerich, Owen. "The Basic Astronomy of Stonehenge." *Technology Review,* December 1977, pp. 64–73.

Gwynne, Peter, and Begley, Sharon. "Once and Future Stars." *Newsweek,* 19 December 1977, pp. 97–100.

Hadingham, Evan. *Circles and Standing Stones.* New York: Walker, 1975.

———. "Approaches to Megalithic Astronomy." Proceedings of Santa Fe Conference on Archaeoastronomy, June 1979.

Hartmann, William K. *Astronomy: The Cosmic Journey.* Belmont, California: Wadsworth, 1978.

Hawkes, Jacquetta. *The Atlas of Early Man.* New York: St. Martin's Press, 1976.

Hawkins, Gerald S. "Amon-Ra, A Sun-pointing Temple." *Preprint,* Smithsonian Astrophysical Observatory, Cambridge, Mass., October 1971.

———. "Ancient Lines in the Peruvian Desert." Final Report to the National Geographic Society. Cambridge, Mass.: Smithsonian Astrophysical Observatory, June 1969.

———. "Astro-Archaeology." Special Report: Number 226, Cambridge, Mass: Smithsonian Astrophysical Observatory, October 28, 1966.

———. "Photogrammetric Survey of Stonehenge and Callanish." National Geographic Society Research Reports: 1965, 1971.

———. *Beyond Stonehenge.* New York: Harper & Row, 1973.

———. "Some Archaeoastronomy Principles." Proceedings of Santa Fe Conference on Archaeoastronomy, June 1979.

———, and White, John B. *Stonehenge Decoded.* Garden City, N.Y.: Doubleday, 1965.

Hicks, Robert D., III. "Astronomy in the Ancient Americas." *Sky and Telescope,* June 1976, pp. 372–77.

Hitching, Francis. *Earth Magic.* New York: Pocket Books, 1978.

Hodson, F. R., ed. "The Place of Astronomy in the Ancient World." Philosophical Transactions of the Royal Society of London, Series A, 276 (Math & Physical Science), 1974.

Hogsett, Vic. "Solstice Watchers of Chaco Canyon." *The Atom,* January 1979, pp. 16–17.

Hoyle, Fred. *On Stonehenge.* San Francisco: W. H. Freeman, 1977.

Hudson, Travis. "California's First Astronomers." Paper presented to the Annual Meeting of American Association for the Advancement of Science. San Francisco: January 1980.

————, and Underlay, Ernest. *Crystals in the Sky: An Intellectual Odyssey Involving Chumash Astronomy, Cosmology and Rock Art.* Anthropological Papers No. 10. Ballena Press/Santa Barbara Museum of Natural History, 1978.

Hutchinson, G. Evelyn. "Long Meg Reconsidered." *American Scientist,* 6(1) [January–February, 1972].

Isbell, William H. "The Prehistoric Ground Drawings of Peru." *Scientific American,* 239(4) [October 1978], pp. 141–53.

————. "Reviews of Three Volumes on Nazca Lines." *Archaeoastronomy Bulletin,* (4) [Fall 1979], pp. 34–40.

Kim, Po Sung. "A 7th-Century Korean Observatory." *Sky and Telescope,* 29(4) [April 1965], pp. 229–30.

Koenig, Seymour H. "Stars, Crescents, and Supernovae in Southwestern Indian Art." *Archaeoastronomy,* 10 (Supplement), 1(1979), pp. 39–50.

Krupp, E. C. "Great Pyramid Astronomy." Los Angeles: *Griffith Observer,* March 1977, pp. 2–18.

————. *In Search of Ancient Astronomies.* Garden City, N.Y.: Doubleday, 1977.

————. "Review: Native American Astronomy." *Sky and Telescope* April 1978, pp. 337–41.

Leakey, Richard, and Lewin, Roger. *Origins.* New York: E. P. Dutton, 1977.

Legesse, A. *Gada: Three Approaches to the Study of African Society.* New York: Free Press, 1973.

Leeming, David. "Review: Hamlet's Mill." *Parabola,* 3(1), pp. 113–15.

Bibliography

Lettvin, Jerome Y. "The Use of Myth." *Technology Review* (June 1976), pp. 52–57.

Lewin, Roger. "An Ancient Cultural Revolution." *New Scientist*, 83(1166) [2 August 1979], pp. 352–55.

Lewis, David. *The Voyaging Stars*. New York: W. W. Norton, 1978.

Linares, Eloy Malaga. *Visita Guiada a Toro Muerto*. Arequipa, Peru: Universidad Nacional San Agustin, 1976.

Love, Thomas. "A Search for an Ancient Explosion in the Southern Sky." *Washington Star* (September 1, 1975).

Lynch, B. M. and Robbins, L. H. "Namoratunga: The First Archaeoastronomical Evidence in Sub-Saharan Africa." *Science*, 200(19 May 1978), pp. 766–68.

McCluskey, Stephen C. "The Astronomy of the Hopi Indians." *Journal for the History of Astronomy*, 8(1977), pp. 174–95.

Marshack, Alexander. *The Roots of Civilization*. New York: McGraw-Hill, 1972.

Mattill, John. "First Line: Woe to the People of Egypt when [a] Comet Appears in Gemini." *Technology Review* (June/July 1978), p. 6.

Mayall, R. Newton and Mayall, Margaret. *Sundials*. Cambridge, Mass.: Sky Publishing, 1973.

Mayer, Dorothy. "Miller's Hypothesis: Some California and Nevada Evidence," *Archaeoastronomy*, 10 (Supplement), 1(1979), pp. 51–74.

"Medicine Wheel Alignments." *Sky and Telescope*, August 1974, p. 77.

Mendelssohn, Kurt. *The Riddle of the Pyramids*. New York: Praeger, 1974.

Michanowsky, George. *The Once and Future Star*. New York: Hawthorn, 1977.

Mitchell, John. *Secrets of the Stones*. New York: Penguin Books, 1977.

Moore, Patrick, and Collins, Pete. *The Astronomy of Southern Africa*. London: Robert Hale and Company, 1977.

Muñoz, Gabriel R. "The Planetarium as an Aid to the Study of Man and His Origin." Proceedings: 6th International Planetarium Directors Conference, Nagoya, Japan, August 1978.

Needham, Joseph, and Ling, Wang. *Science and Civilization in China.* Vol. 3: *Mathematics and the Sciences of the Heavens and the Earth.* Cambridge, England: Cambridge University Press, 1959.

Neugebauer, Otto. *The Exact Sciences in Antiquity.* 2nd ed. Providence: Brown University, 1957.

Newham, C. A. *The Astronomical Significance of Stonehenge.* Leeds, England: John Blackburn, 1972.

Nuttall, Zelia. "The Astronomical Methods of the Ancient Mexicans." Boas Anniversary Volume. New York: Stechert, 1906.

Oberg, James. "Strange Science Bedfellows: Ancient Eclipses and Modern Astronomy." *Griffith Observer,* February 1979, pp. 2–10.

Patrick, Jon. "A Reassessment of the Lunar Observatory Hypothesis for the Kilmartin Stones." *Archaeoastronomy,* 10 (Supplement), 1(1979), pp. 78–85.

Peng-Yoke, Ho; Parr, F. W., and Parsons, P. W. "The Chinese Guest Star of A.D. 1054 and the Crab Nebula." *Vistas in Astronomy.* A. Beer, ed., Vol. 13. Oxford: Pergamon, 1974.

Ponting, Gerald, and Ponting, Margaret. *The Standing Stones of Callanish.* Stornoway, Scotland: Privately published, 1977.

Rensberger, Boyce. "Early, Earlier, and Earliest Man." *The New York Times,* 26 June 1977, p. E-11.

––––––. "Rock Art Shows a Supernova." *The New York Times,* 10 September 1975, p. 39.

––––––. "Prehistoric Astronomy Was Pretty Good Science." *The New York Times,* 19 February 1978, p. E-12.

––––––. "The World's Oldest Works of Art." *The New York Times Magazine,* 21 May 1978, pp. 26–42.

Reyman, Jonathan E. "Archaeoastronomy and the History of Astronomy." *Archaeoastronomy* (Bulletin), 2(2) [Spring 1979], pp. 11–13.

––––––. "Astronomy, Architecture, and Adaptation at Pueblo Bonito." *Science,* 193(4257), pp. 957–62.

"Rock Glyphs and A.D. 1054 Supernova," *Science News,* 115(27) [7 July 1979].

Rodriguez, Luis F. "Ancient Astronomy in Mexico and Central America." *Mercury,* January/February 1975, pp. 25–27.

Index